Effective Teaching in Large STEM Classes

Online at: https://doi.org/10.1088/978-0-7503-5231-4

IOP Series in Physics Education

The IOP Series in Physics Education aims to provide comprehensive, authoritative and innovative coverage for those that teach physics and related subjects at universities and other higher and further education institutions, and for those involved in physics education research.

Series Editor
Professor Peter Main
King's College London, UK

About the Editor
Peter Main obtained his PhD from the University of Manchester and, after post-docs in Manchester and Helsinki, he joined the University of Nottingham as a Lecturer in Physics in 1979. Following promotions to Reader and Professor, he eventually became Head of the School of Physics and Astronomy. His principal research interests were in quantum fluids and quantum transport in semiconductor and metallic heterostructures. He was also involved in many teaching innovations.

In 2002 he left Nottingham to join the Institute of Physics as Director of Education and Science. In this post he had overall responsibility for the Institute's work in education at all age levels, research, and diversity. Among many projects, he worked closely with Ofqual and awarding bodies on curriculum matters and with government to increase the number of physics teachers. He also initiated several projects improving the diversity of participation in physics.

In 2015 he joined King's College to become Head of Physics; he retains his interest in many projects in physics education and diversity.

About the Series
The IOP Series in Physics Education aims to provide comprehensive, authoritative, and innovative coverage for those that teach physics and related subjects at universities and other higher and further education institutions, and for those involved in physics education research.

The series supports evidence-informed professional practice and will cover topics including the following: assessment methods; feedback; conceptual understanding; problem solving; teaching methods; education technology; pedagogical theory; curriculum design; student engagement; misconceptions; employability; and social aspects of education.

Authors are encouraged to take advantage of electronic publication through the use of colour, animations, video, data files, and interactive elements, all of which offer particular benefits in communicating pedagogy.

Do you have an idea for a book you'd like to explore?
We are currently commissioning for the series; if you are interested in writing or editing a book please contact Caroline Mitchell at caroline.mitchell@ioppublishing.org.

A full list of titles published in this series can be found here: https://iopscience.iop.org/bookListInfo/iop-series-in-physics-education

Effective Teaching in Large STEM Classes

Edited by
Anna K Wood
School of Mathematics, University of Edinburgh, Edinburgh, UK

IOP Publishing, Bristol, UK

© IOP Publishing Ltd 2023

All rights reserved. No part of this publication may be reproduced, stored in a retrieval system or transmitted in any form or by any means, electronic, mechanical, photocopying, recording or otherwise, without the prior permission of the publisher, or as expressly permitted by law or under terms agreed with the appropriate rights organization. Multiple copying is permitted in accordance with the terms of licences issued by the Copyright Licensing Agency, the Copyright Clearance Centre and other reproduction rights organizations.

Permission to make use of IOP Publishing content other than as set out above may be sought at permissions@ioppublishing.org.

Anna K Wood has asserted her right to be identified as the editor of this work in accordance with sections 77 and 78 of the Copyright, Designs and Patents Act 1988.

ISBN 978-0-7503-5231-4 (ebook)
ISBN 978-0-7503-5228-4 (print)
ISBN 978-0-7503-5232-1 (myPrint)
ISBN 978-0-7503-5230-7 (mobi)

DOI 10.1088/978-0-7503-5231-4

Version: 20230801

IOP ebooks

British Library Cataloguing-in-Publication Data: A catalogue record for this book is available from the British Library.

Published by IOP Publishing, wholly owned by The Institute of Physics, London

IOP Publishing, No.2 The Distillery, Glassfields, Avon Street, Bristol, BS2 0GR, UK

US Office: IOP Publishing, Inc., 190 North Independence Mall West, Suite 601, Philadelphia, PA 19106, USA

Contents

Preface	xii
Acknowledgements	xv
Editor biography	xvi
List of contributors	xvii
Contributor biographies	xviii

1 Large classes in STEM education—potential and challenges   1-1
Anna K Wood and Nicolas Labrosse

1.1	Introduction	1-1
1.2	Definition of large classes	1-2
1.3	Why do we have large classes?	1-3
1.4	Should we have large classes?	1-4
1.5	Challenges of large classes	1-5
	1.5.1 Opportunities for interactions	1-5
	1.5.2 Anonymity	1-6
	1.5.3 Heterogeneous background of students	1-6
	1.5.4 Assessment and feedback	1-7
1.6	Benefits of large classes	1-7
	1.6.1 Social aspects	1-7
	1.6.2 Diversity	1-8
1.7	Approaches to making large classes successful	1-8
1.8	Conclusion	1-9
	References	1-10

2 Active learning in large classes   2-1
Anna K Wood

2.1	What is active learning?	2-1
2.2	Flipped classroom	2-2
2.3	Active learning versus direct instruction	2-3
2.4	Characterizing active learning approaches—ICAP	2-4
2.5	Evidence for active learning	2-4
2.6	Barriers to implementing active learning	2-5
2.7	Theoretical basis for active learning	2-7
	2.7.1 Cognitive constructivism	2-7
	2.7.2 Transactional distance theory	2-8

		2.7.3 Discourse and dialogue	2-10
2.8	Active learning in practice—focus on Peer Instruction		2-10
		2.8.1 Peer Instruction	2-11
		2.8.2 The role of the teacher talking	2-11
		2.8.3 Student–teacher interactions	2-12
		2.8.4 Student–student interactions	2-13
2.9	Conclusions		2-14
	References		2-15

3 Practical approaches to active learning   3-1
Alison Voice

3.1	Features of active learning in the classroom	3-1
3.2	Question types	3-3
	3.2.1 Designing questions	3-4
3.3	Methods of deploying active learning in large classes	3-8
	3.3.1 Flipped classroom	3-9
	3.3.2 Integrated teaching and active learning	3-10
	3.3.3 Active learning workshops	3-10
3.4	Room layout and group formation	3-11
3.5	Learning preferences	3-13
3.6	Concluding remarks	3-14
	References	3-15

4 Using classroom observation tools to characterize large classes   4-1
Anna K Wood and George Kinnear

4.1	What are classroom observation tools? Why do we need them?	4-1
4.2	Summary of characterization tools	4-2
	4.2.1 TDOP	4-2
	4.2.2 COPUS	4-3
	4.2.3 PORTAAL	4-3
	4.2.4 FILL/FILL+	4-3
	4.2.5 DART	4-4
	4.2.6 Other tools	4-4
4.3	Comparison of tools	4-4
4.4	Guide to using FILL+ (and interpreting the data)	4-5
4.5	Coding with FILL+	4-7
4.6	Understanding FILL+ data	4-8

4.7	Examples of how classroom characterization tools are used in practice	4-9
	4.7.1 Collecting data on faculty practices	4-9
	4.7.2 Classifying the types of teaching approach used	4-11
	4.7.3 Gaining greater understanding of pedagogical approaches	4-11
	4.7.4 Supporting individual teachers to make changes	4-12
4.8	Future potential	4-13
	4.8.1 Peer observation	4-13
	4.8.2 Supporting teacher change	4-14
	4.8.3 Supporting professional development programmes	4-14
	4.8.4 Tracking longitudinal changes to teaching practices	4-14
4.9	Conclusion	4-14
	References	4-15

5   Authentic and inclusive (summative) assessments   5-1
Firas Moosvi and Simon Bates

5.1	Introduction	5-1
	5.1.1 Purpose of assessment	5-3
	5.1.2 Aspirational goal for assessments	5-3
5.2	General strategies	5-4
	5.2.1 Frequent testing	5-5
	5.2.2 Leveraging learning technologies	5-6
	5.2.3 Transforming assessments with inclusive policies and pedagogies	5-8
	5.2.4 Alternative grading paradigms	5-9
5.3	Specific implementations	5-10
	5.3.1 Collaborative two-stage final exams	5-11
	5.3.2 Reflections on exams	5-11
	5.3.3 Automated feedback during final exams	5-13
	5.3.4 Oral examinations	5-13
	5.3.5 Experiential learning	5-14
5.4	Conclusion	5-15
	References	5-16

6   Computer-marked assessment and concept inventories   6-1
Sally Jordan

6.1	Introduction	6-1
6.2	Definitions and history	6-2
6.3	Advantages and disadvantages of computer-marked assessment	6-3

	6.3.1 Why use computer-marked assessment?	6-3
	6.3.2 Why not?	6-5
	6.3.3 Disadvantage or advantage?	6-6
6.4	Assessment design and integration with teaching	6-7
	6.4.1 Motivation and purpose	6-7
	6.4.2 The context in which the assessment is used	6-8
	6.4.3 How does the assessment run?	6-9
	6.4.4 Accessibility	6-11
6.5	Question types	6-11
	6.5.1 Selected-response questions	6-12
	6.5.2 Simple constructed-response questions	6-12
	6.5.3 Questions based on computer-algebra	6-13
	6.5.4 Assessing words, phrases and essays	6-13
	6.5.5 More advanced question types	6-15
6.6	Writing questions and feedback	6-15
	6.6.1 Writing questions	6-15
	6.6.2 Distractors, correct and incorrect answers and feedback	6-16
	6.6.3 Evaluation and iterative design	6-17
6.7	How far should we go?	6-18
	References	6-18

7 Case study 1: an introductory physics course — 7-1
Ross K Galloway

7.1	General context of the course	7-1
7.2	Structure of the course	7-2
	7.2.1 Personal reading	7-2
	7.2.2 Peer instruction lectures	7-4
	7.2.3 Workshops	7-7
	7.2.4 Overall structure of the course	7-9
7.3	Assessment	7-10
7.4	Evaluation	7-10
	7.4.1 Peer instruction episodes	7-11
	7.4.2 Force concept inventory	7-12
	7.4.3 Student perspectives	7-13
7.5	Teacher perspectives	7-14

7.6	Summary	7-15
	Acknowledgements	7-15
	References	7-15

8 Case study 2: tailored active blended learning in a foundation year chemistry module — 8-1
Simon J Lancaster, Daniel Elford and Eleanor Gill

8.1	Introduction	8-1
	8.1.1 The context	8-1
	8.1.2 Peer instruction in chemistry	8-2
8.2	Tailored active blended learning	8-2
	8.2.1 The rationale	8-2
	8.2.2 A weekly structure	8-3
	8.2.3 Introductory quiz	8-3
	8.2.4 Chunked videos	8-4
	8.2.5 Synchronous teaching	8-4
	8.2.6 Formative test	8-6
	8.2.7 Workshops	8-6
8.3	Class-sourcing misconceptions	8-7
8.4	Fostering and utilizing sub-cohort identities through gamification	8-8
	8.4.1 Extending in-person peer discussion with online tools	8-8
	8.4.2 Extrinsic motivation through badging and gamification	8-9
	8.4.3 Enhancing peer instruction through gamification	8-9
8.5	Concluding remarks and advice	8-10
8.6	Future direction	8-11
	Acknowledgements	8-11
	References	8-12

9 Case study 3: an introductory linear algebra course — 9-1
Pamela Docherty

9.1	Introduction	9-1
	9.1.1 Context	9-1
	9.1.2 History	9-2
9.2	Course structure	9-2
	9.2.1 Overview	9-2
	9.2.2 Pre-class activities	9-2
	9.2.3 Active learning lectures	9-3

	9.2.4 Workshops	9-5
	9.2.5 Assessment structure	9-6
9.3	Evaluation	9-8
	9.3.1 What didn't work, changes post-COVID-19	9-8
	Acknowledgements	9-9
	References	9-9

10 Case study 4: personalised learning by student-posed questions during biology lectures   10-1
Heather McQueen

10.1	Are students talking in your class? Should they be?	10-1
10.2	How can we encourage personally relevant learning for every student during large class teaching?	10-2
10.3	How was the quecture strategy received?	10-5
10.4	Can the quecture strategy really make a difference: potential benefits?	10-7
10.5	Can the quecture strategy really make a difference: potential pitfalls?	10-8
10.6	How can the quecture strategy be exploited when lectures have moved online?	10-9
10.7	How can you use quecture questions in your large classes?	10-9
	References	10-10

11 Case study 5: the learning assistant model for engaging students   11-1
Valerie K Otero

11.1	Introduction	11-1
11.2	Research on the effectiveness of the LA model throughout the US	11-5
11.3	Research that suggests reasons for why the LA model is effective	11-7
11.4	LAs are used in different contexts	11-8
11.5	A model of institutional change	11-11
11.6	Get involved: the International LA Alliance	11-12
11.7	LA Alliance: https://learningassistantalliance.org/	11-12
	11.7.1 LA program at CU Boulder: https://www.colorado.edu/program/learningassistant/	11-12
11.8	Final thoughts	11-12
	References	11-15

12	**Effective teaching in large classes; looking through and beyond the COVID-19 pandemic**	**12-1**
	Simon Bates and Firas Moosvi	
12.1	Introduction	12-1
12.2	The emerging post-COVID-19 context	12-2
12.3	Course design and logistics	12-3
12.4	Course content	12-4
12.5	Interactions	12-5
12.6	Assessments	12-6
12.7	Conclusion and outlook	12-8
	References	12-9

Preface

I was fortunate enough to be a physics undergraduate at the University of Birmingham in the early 1970s. Although much of the programme was traditional, with formal lectures and unseen examinations, there was a strand that might loosely be construed as encouraging thinking. This culminated in the 6-week *Group Studies*, which took place after the final examinations and which comprised essays, problem solving, oral presentations and an extended project. The exercise had a profound effect on my outlook: first, it was, by some way, the hardest I had ever worked as a student but equally the most satisfying. Over a longer timescale, it influenced the way I have taught, with the principle that the best way to learn was to do so actively.

This book is part of a collection, of which I am series editor, of texts on higher education in physics and related subjects. The series, and this book in particular, have the primary purpose of appealing to academics who are interested in making their teaching more effective but perhaps do not have the time to find out how to do so. Consequently, although the material provided is soundly research-based, it is presented in terms of real activities and case studies, highlighting the advantages and disadvantages of each approach.

Physics, and STEM (science, technology, engineering and maths) in general, present a special challenge for educators. There is an extensive knowledge base coupled with significant conceptual demands. With large classes the norm, a high proportion of student activity has frequently involved passive attendance at lectures. This book addresses head-on the task of creating a more active learning environment, while preserving what some would see as necessary content and, crucially, not involving a massive increase in staff workload.

In the first chapter, Anna Wood and Nicolas Labrosse provide an introduction to the issues associated with teaching large classes, including a discussion of what we mean by a large class. Rather than simply defining a threshold number of students, they consider the student experience, that is, whether there are strong student–teacher and student–student interactions or simply the type of information transfer we associate with traditional, large-class lectures. They set the scene for the later chapters, identifying the conditions for successful learning, highlighting the importance of regular interactions, active learning, managing expectations and, increasingly, the use of online resources.

In chapter 2, Anna Wood makes a persuasive case for active learning, providing strong evidence that it does indeed improve students' learning, both in terms of depth and longevity. While providing a discussion on the theoretical reasons for this improvement, she also discusses more pragmatic issues, such as the importance of staff commitment and how to overcome initial student resistance. In the UK, the latter point is vital because student perceptions play a large role in how teaching quality is viewed externally. An important observation is that it is the active learning itself which is pedagogically important, not necessarily the structure through which it is delivered. A flipped classroom does not in itself lead to active learning, while a session involving direct instruction does not have to result in a passive student experience. What matters is that the students are required to think and, above all, talk to each other about the subject. As any teacher knows, by explaining we soon find out what we do not understand.

With chapter 2 providing the reasoning, chapter 3, written by Alison Voice, delivers the nuts-and-bolts information on how to make it happen. She explains how to prepare the students, who might be used to a standard diet of lectures, to participate in the more active approach and even goes into detail on what may seem at first sight to be relatively trivial matters, such as the layout of the room, a matter which I know to my own cost, can make a huge difference. She also considers the very important issue of how to accommodate neurodiverse students who, for example, might find it difficult to interact easily with their colleagues.

For many academics, moving away from standard lectures may constitute a real challenge, not so much in terms of preparation time—in practice, a shift to a more active approach may be undertaken in a piecemeal manner—but in terms of unfamiliarity. Evidently, any student-active approach is going to involve some elements of students talking to each other but, for example, is it better for them to do that cold, or do they learn more effectively if they have a lecture beforehand? Anna Wood and George Kinnear describe several classroom observation tools that provide information on just such questions. They concentrate on the FILL (framework for interactive learning in lectures) tool and how it can be used to obtain a better understanding of pedagogical approaches. Not every teacher will have the time or inclination to go so deeply into the evaluation process but these tools form an excellent basis for professional team development.

Assessment plays a part in student learning. It drives student behaviour—one often finds students reluctant to engage with 'what doesn't count'—and it also indicates which activities are deemed important. And formative assessment, with feedback, is an essential element of the learning process. Firas Moosvi and Simon Bates examine in detail the various assessments that can be used in a university environment, in passing pointing out that 'traditional' examinations, despite their ubiquity, do not have an unblemished history. They examine the various strategies that can lead to authentic assessments, that is, those matched closely to the learning outcomes of the teaching. Finally, they propose a scheme in which summative assessment is gradually replaced by formative assessment linked closely to the teaching, while recognising that, with current workloads, such an approach may not be plausible.

Assessment is also the theme of chapter 6 by Sally Jordan, who tries to address the task of lightening the staff burden by employing computer marking. This is an area which is likely to become much more sophisticated over the coming years. Evidently, computers have been used to mark multiple choice questions for many decades. While such questions are still used to test conceptual understanding—not least in the concept inventories which are also discussed in this chapter—much more exciting is the possibility of marking free-text answers and supplying relevant feedback and tutorial material for incorrect responses.

Several chapters are dedicated to case studies, across different STEM subjects. Three STEM subjects at University of Edinburgh: physics (Ross Galloway); biology

(Heather McQueen) and mathematics (Pamela Docherty) are represented, as well as chemistry at UEA (Simon Lancaster, Daniel Elford and Eleanor Gill) and physics in the USA context (Valerie Otero). The studies highlight many different approaches but each demonstrates the efficacy of active learning. Heather McQueen tackles the thorny issue of how to match learning to the individual in a large class. Her 'quecture' approach to the flipped classroom uses student-posed questions based on what they find difficult instead of a one-size-fits-all set of teacher-set problems.

The UEA contribution describes an introductory chemistry module for students with a very broad range of ability; the task is to avoid bewildering those students who have almost no chemistry pre-knowledge while not boring those that do. The authors draw heavily on peer instruction, extended by technology, again concentrating on a personalized approach to learning. Ross Galloway describes a similar introductory course, this one in physics involving a huge class of more than 300. Again, the basic idea involves the flipped classroom. Because the course has been running for more than a decade, Ross is able to provide detailed information, not just about the day-to-day activities but also about their efficacy. I am particularly struck by how, initially, students often struggle to see the difference between learning and being taught but, afterwards, they much prefer the active approach.

Following the success in the Physics Department, Pamela Docherty, from Mathematics, describes how she and colleagues have used active learning for a massive class of around 900 students learning linear algebra. Although the project is clearly a success, Pamela is not afraid to discuss some of the difficulties encountered, for example, whether to distribute problem sheets in advance of classes. She also accepts that some students simply do not wish to discuss questions with their contemporaries. But what is clear is that, whatever issues there are with the active learning approach, they are nothing compared with those encountered with the traditional lecture/exam methodology. A pleasing side-effect is how the more interactive style of teaching has led to an improved atmosphere in the department as a whole, an outcome that chimes with my own experience.

In a case study from the USA, Valerie K Otero from the University of Colorado Boulder contributes a thoughtful chapter on the use of learning assistants, who are UG students themselves, to help teach students in the earlier years. I am personally drawn to that idea since I feel that there is a twofold gain: the junior students are less nervous about asking questions than they would be with the Professors but, in addition, the senior students also deepen their understanding of basic material by the simple act of explaining it.

At the time of writing, the COVID pandemic is a recent, dismal memory. However, the cloud had a silver lining. With crowded lectures and examination halls precluded by lockdown, academics were forced to develop more active approaches to learning and assessment, some, it has to be said, with less enthusiasm than others. Nevertheless, the change occurred and in the final chapter, Simon Bates and Firas Moosvi review that change and discuss its possible legacy, given that so many innovative resources were developed. It provides a fitting end to a challenging book, which I hope will be a useful resource in STEM classrooms around the world.

Acknowledgements

Firstly, a huge thank you to all the contributors in this volume who have shared their wisdom and expertise so generously. I would also like to thank everyone on the MSc in E-learning at the University of Edinburgh, especially Clara O'Shea and Christine Sinclair whose support and encouragement enabled me to start an unexpected new career in education research. Particular thanks to Hamish Macloed without whom I may have never discovered that science education was a field of research and for connecting me to the Physics Education Research (PER) Group in Edinburgh. Thank you to Ross Galloway and Judy Hardy of the Edinburgh PER group for so many fruitful collaborations and deep discussions about teaching and learning and to my colleagues in the School of Mathematics, particularly George Kinnear and Chris Sangwin for supporting my efforts to edit and contribute to this book. Finally thank you to Peter Main for asking me (and trusting me) to edit this volume and to everyone at the IOPP for bringing it to life.

Editor biography

Anna K Wood

Anna K Wood has a PhD in Physics (Durham University, 2000) and an MSc in E-learning (University of Edinburgh, 2013). She is currently a researcher working in the School of Mathematics at the University of Edinburgh. Her research interests include understanding the role that technology plays in large STEM classes, how dialogue leads to learning and how seeing qualitative data about the activities that take place in a lecture can help teachers to reflect on their teaching. In 2016 she developed the FILL (framework for interactive learning in lectures) protocol for characterising activities in lectures.

List of contributors

Simon Bates
Department of Physics and Astronomy, University of British Columbia, Canada

Pamela Docherty
School of Mathematical and Computer Sciences, Heriot-Watt University, UK

Daniel Elford
School of Chemistry, University of East Anglia, UK

Ross K Galloway
School of Physics and Astronomy University of Edinburgh, UK

Eleanor Gill
School of Chemistry, University of East Anglia, UK

Sally Jordan
School of Physical Sciences, The Open University, UK

George Kinnear
School of Mathematics, University of Edinburgh, UK

Nicolas Labrosse
School of Physics and Astronomy, University of Glasgow, UK

Simon Lancaster
School of Chemistry, University of East Anglia, UK

Heather McQueen
Institute of Cell Biology, University of Edinburgh, UK

Firas Moosvi
Department of Computer Science, University of British Columbia, Canada

Valerie Otero
School of Education, University of Colorado, Boulder, Co, USA

Alison Voice
School of Physics and Astronomy, University of Leeds, UK

Anna K Wood
School of Mathematics, University of Edinburgh, UK

Contributor biographies

Simon Bates
Simon Bates is Vice Provost and Associate Vice-President Teaching and Learning at UBC, and Professor of Teaching in the Department of Physics and Astronomy. His research interests span technology enhanced learning, physics education research and fostering cultures of leadership and innovation in teaching and learning higher education.

Pamela Docherty
Pamela Docherty is an Assistant Professor of Mathematics at Heriot-Watt University's School of Mathematical and Computer Sciences in Edinburgh, UK, where she leads the Technology Enhanced Mathematical Sciences Education research group. Her interests include active teaching methods, assessment and caring pedagogic approaches.

Daniel Elford
Daniel Elford has a first-class honours degree in chemistry from Kingston University London (2014) an MSc in Information Technology at Anglia Ruskin University (2018). He is a PhD student in the School of Chemistry at the University of East Anglia focussing on integration of augmented and virtual reality technologies into chemistry higher education.

Ross Galloway
Ross Galloway is a Senior Lecturer in the School of Physics and Astronomy at the University of Edinburgh. He teaches on the undergraduate programmes in physics and astronomy and also conducts pedagogic research as a member of the Edinburgh Physics Education Research group (EdPER). His research interests include the development of student problem solving skills, diagnostic testing, active learning and flipped classroom pedagogies.

Eleanor Gill
Eleanor Gill graduated from the University of East Anglia in 2021 with a first-class master's degree in Chemistry. Since 2021, she has been working as a Nuclear Chemistry Technician for EDF.

Sally Jordan
Sally Jordan is Professor of Physics Education at the UK Open University (OU). She has extensive experience of teaching large OU classes in physics, interdisciplinary science and mathematics. She was the first person at the OU to use interactive online computer-marked assessment in her teaching. Her research interests include demographic outcome gaps in physics, authentic remote experimentation, the use of concept inventories and the impact of assessment on students.

George Kinnear
George Kinnear is a Reader in the School of Mathematics at the University of Edinburgh. His research interests are in effective uses of technology to support undergraduate mathematics teaching and learning.

Nicolas Labrosse
Nicolas Labrosse is Senior Lecturer in the School of Physics and Astronomy at the University of Glasgow, where he co-chairs the Astronomy and Physics Education group. His interests include students transitions, sense of belonging and students' identity as physicists. His current research focuses on assessments.

Simon J Lancaster
Professor Simon J Lancaster has a Chair in Chemistry Education and is currently Associate Dean for Learning and Teaching in the Science Faculty at The University of East Anglia. Simon is probably best known for his promotion of active learning to add value to the lecture theatre experience.

Heather McQueen
Heather McQueen is Professor of Biology Education at the University of Edinburgh. Her teaching aims to foster engagement and deeper learning through interactivity and specifically using students' questions for personalized learning, which has also been the subject of her research.

Firas Moosvi
Firas Moosvi is a Lecturer at the University of British Columbia and teaches large classes in Physics, Computer Science, and Data Science. His research interests include alternative grading paradigms, building inclusive classrooms, and learning technologies.

Valerie Otero
Valerie Otero is a Professor of STEM Education at the University of Colorado Boulder, specializing in physics education research. She is co-founder and faculty director of the Learning Assistant (LA) program, the International LA Alliance, and PEER Physics high school curricular materials.

Alison Voice
Alison Voice is a Professor of physics education in the School of Physics and Astronomy at the University of Leeds. She has many years' experience of engaging students in active learning, to deepen their understanding of the discipline and enhance peer collaboration in problem solving.

IOP Publishing

Effective Teaching in Large STEM Classes

Anna K Wood

Chapter 1

Large classes in STEM education—potential and challenges

Anna K Wood and Nicolas Labrosse

In this chapter we explore the potential and challenges of large classes in STEM teaching. We begin by attempting to define a large class, then explore the reasons why they exist and why they are likely to continue for the foreseeable future. We then discuss the specific problems associated with large classes such as the difficulty in creating interactions and the danger of students feeling anonymous, and point to some of the solutions to these problems, such as active learning techniques, which will be discussed in more detail throughout this book.

1.1 Introduction

Although a universal phenomenon in higher education, large classes have typically been seen as detrimental to student learning (Ehrenberg *et al* 2001, Cuseo 2007). The reasons for this include a belief that they are associated with weak student engagement and fewer opportunities for interactions with the teacher and with each other. Certainly large classes that are not carefully designed may create a 'transactional distance' (see chapter 2), that is, a psychological and communications space between student and teacher which hinders learning, leading them to be experienced as impersonal, with a sense of isolation and anonymity. However, others believe that with the use of active learning pedagogies, large classes can provide stimulating learning environments which lead to critical thinking and deep learning (Hornsby and Osman 2014). Despite the worry about the potential downsides of large classes, undergraduate science classes in the UK commonly have anywhere between 50 and 300 (or more) students, and in early and foundational years, numbers are usually at the upper end of this range. This situation is not likely to change as the move towards the massification of higher education has strong financial and political (Hornsby and Osman 2014) drivers.

In this chapter, as in the rest of this volume, we purposely use the term 'large class' rather than 'lecture' to discuss the learning environment encountered by STEM students. This is because a 'lecture' conjures up a particular type of pedagogy—specifically one in which a person (the lecturer) talks for the majority of the time, and students listen and take notes. In contrast the term 'large class' is free from any pedagogical preconceptions and may, therefore, take any number of forms, many of which are discussed in later chapters. Similarly the term 'teacher' is used in place of 'lecturer'.

Although there are huge challenges associated with providing high quality learning in large classes, it is our belief, as evidenced by the contributions in this volume, that it is possible, through the introduction of research informed active learning approaches, to create a good quality, stimulating, learning environment which leads to deep learning, even in STEM classes with diverse cohorts and large numbers of students. This chapter will explore how to define large classes, and discuss their challenges and potential benefits.

1.2 Definition of large classes

There is no agreed definition of a large class although various different sizes have been proposed; Winke and Rawal (2018) suggest a large class is 35 or more students and Cuseo (2007) reports it is common for classes of over 50 to be seen as large. In contrast Jungic *et al* (2006) define a large class as having more than 350 students and Cash *et al* (2017) found from a survey of students and teachers that a class was felt to be large at around 240 students. Rather than defining a large class by a numerical threshold, a more useful measure would be to consider how the number of students impacts on their experience of the class (Hornsby *et al* 2013). Cash *et al* report that students described a large class as one that is impersonal and anonymous and which has an atmosphere 'where neither their peers nor their instructor noticed whether they were absent or attentive' (Cash *et al* 2017). In contrast the students described a small class as a more personal learning experience which created a sense of community and where the pace of the learning could be adjusted in response to students' needs.

The factors which influence how large or small a class feels include the teaching strategies employed, how much interaction takes place, and the extent to which students feel actively engaged with their learning. Another important aspect which affects the feel of the class is the teacher to student ratio. In the UK it is common for a large class to be taught by a single teacher alone whereas in other countries such as the US the teacher is supported by teaching assistants, giving a higher teacher to student ratio. This latter scenario is more likely to make students feel visible and involved, indeed Lenton (2015) found that the staff student–ratio is a predictor of student satisfaction.

The teaching strategies employed also make a significant difference to how large or small the class feels, however it is important to note that smaller classes often use similar pedagogical approaches to larger classes. For example there is evidence from a survey of 2000 courses in North America that smaller class sizes do not necessarily result in an increase in instructional innovation and student-centred approaches (Stains *et al* 2018). This may be because teachers do not change their pedagogical approach when moving from large to small classes (Wright *et al* 2019). This can happen for a variety of reasons such as the demands of the physical space, highly structured curricula, and lack of pedagogical skills to facilitate discussions.

However, the size of the class does influence how much effort is required to create good quality interactions. For example, with a class of 20 students it is much more straightforward to generate whole class discussions where real dialogue takes place between the teacher and the students, and where most students have the opportunity to take an active part. In larger classes students can be put off from speaking in front of their peers, whereas in smaller classes students are likely to be familiar with each other and therefore more comfortable adding their voice to the discussion.

In contrast, large classes are often taught through direct instruction, where the teacher uses a continuous monologue to introduce new ideas while the students listen and make notes, with very little interaction. It is much harder to introduce whole class discussions in this context, and when that is attempted, most of the students are not directly involved in the dialogue. This style of teaching is likely to be experienced as a large class regardless of the number of students in the class. Indeed Cash *et al* (2017) found that 42% of student respondents felt that a numerically large class can feel small if the instructor was engaging, if they made them feel like individuals and if they incorporated small-group activities which enabled them to connect with their peers.

1.3 Why do we have large classes?

One of the key drivers of large classes in post-secondary STEM education is the 'massification' of higher education (Hornsby and Osman 2014), a term used to describe the rapid increase in student numbers that took place at the end of the 20th century, and which has continued into the 21st century (Scott 1995). Massification is a global phenomenon, purposely pursued with the aim of improving health, empowerment and economic development (Bloom *et al* 2006, Santiago 2008). It has been argued that increasing enrolment in higher education is a consequence of an increased sense of social justice and fits with the democratization of education (Altbach 1992). Massification has led to increased class sizes, but it is worth pointing out that this is not a necessary outcome, rather, a consequence of the lack of a corresponding increase in financial (and therefore human) support from public sources in the higher education sector (Mohamedbhai 2008). The increase in class sizes has resulted in concerns that academic standards will slip, as, it is argued, while large classes are suited to information transfer, they are not suited to the acquisition of disciplinary knowledge, because this requires contact between lecturers and students, which is limited in large classes (Allais 2014). Yet some researchers, such as Arvanitakis (2014) discuss the benefits of massification, arguing that it offers the potential to 'promote the emergence of citizen scholars in sections of the population that have for far too long been ignored', and that the challenges of large classes can be overcome by pedagogical approaches.

1.4 Should we have large classes?

While large classes are likely to be a feature for at least the foreseeable future, it is worth considering the arguments both for and against their use. One of the most influential works concerning the utility of lectures is 'What's the Use of Lectures' by Bligh (1998). One quote in particular notes that, while the lecture 'may be used appropriately to convey information ... it cannot be used effectively on its own to promote thought or to change and develop attitudes without variations in the usual techniques' (p 13). This quote is often used as an argument to show that there is no value in the lecture as it does not encourage thought or inspire students. However, in reality, Bligh's discussion is more nuanced than this: rather than arguing against all lectures, he is instead saying that the lecture should be given less prominence in the curriculum, and that other learning experiences should be seen as important. Indeed as we will see in this volume, modern STEM courses will often combine lectures with workshops, tutorials and laboratory experiments to provide a holistic and varied learning experience. Furthermore, the final part of the quote 'variations in the usual techniques' is particularly relevant, implying that techniques can be employed which *will* promote thought, change attitudes and support the development of critical thinking. These techniques are a key feature of this volume.

However, discussions in the popular press have recently been particularly polarized with a strong movement both for and against the use of lectures, for example articles discussing how the 'death of the lecture' is imminent (e.g., *Campus Review* 2019), are counterbalanced by articles singing the praise of lectures (e.g., *The Guardian* 2013).

The arguments against the lecture are mainly that they do not engage students, that by a few weeks into the semester, students no longer attend, that students who do come have a short attention span, and rather than listening to the lecturer, are likely to be using social media, and that the teaching method of 'stand and deliver' is not effective. In contrast those in favour of lectures cite the way that they provide meaning about the subject, synthesize important ideas, communicate passion and enthusiasm, create a community of learners and a common intellectual experience, demonstrate expert-like thinking and problem-solving, and provide academics with a new way of thinking about the subject. What pieces such as these tend to do however, is to conflate two things: class size and pedagogical approach. As we have discussed, large classes are likely to be here to stay, therefore if we can't change the number of students, we can at least change the pedagogical approach. Good pedagogical approaches can overcome the challenges cited, while also providing the advantages of large classes, such as creating a community and common experience for students (Lawrence 2022).

Although we might assume that large classes are always highly detrimental to students' learning, the literature on the subject is mixed, with some research showing a large negative impact on grades and high levels of attrition (Cuseo 2007, Kara *et al* 2021), others finding more minimal effects (Ake-Little *et al* 2020) and some demonstrating no negative effect at all on student learning and achievement (Gleason 2012). For example in a large study comparing STEM and non-STEM classes, STEM classes were found to have the biggest negative effect on grades with respect to class size (Kara *et al* 2021) and Cuseo (2007) reports that the drop-out rate for first year students was more than 25% at institutions with four year courses. In contrast, Gleason (2012) compared mathematics classes with around 30 students to those with over 100 students, with a similar curriculum, and found very little difference in outcomes, and a small positive effect for student satisfaction in the large classes. Gleason hypothesized that these findings were due to the extensive use of technology both in and out of class. Care needs to be taken when using student perceptions, as students tend to assign higher ratings to instructors and courses when the class size is smaller, even when accounting for the pedagogical approach (Wright *et al* 2019).

The conclusion from the literature is that large classes **can** be detrimental to student learning, achievement and completion rates, but that this doesn't need to be the case. With carefully thought out student-centred pedagogical approaches large classes can be just as successful as smaller classes and should not be feared as being second best.

1.5 Challenges of large classes

1.5.1 Opportunities for interactions

Although we believe that large classes have the potential to become good quality learning environments, we also acknowledge that this isn't easy to achieve and some key challenges need to be addressed. If we are to take a (social) constructivist view of learning, in which learners construct knowledge, rather than receive it passively, and that the social context in which this happens is important, then the most pressing challenge in large classes is the quality and quantity of interactions that students experience. Research shows that both student–teacher and student–student interactions are critical to the success of large classes (Jerez *et al* 2021) and that they support students to develop critical thinking, overturn misconceptions and lead to deep learning (Driver *et al* 1994). Yet the commonly used pedagogical approach in large classes, in which the teacher delivers a monologue while students listen and take notes, involves very few such interactions. Work by Stains *et al* (2018) in the US and Kinnear *et al* (2021) in the UK using classroom observation tools such as COPUS and FILL+ shows that it is not uncommon for students in some classes to spend as little as 5% of the time on any type of interaction. This has multiple adverse consequences for students' learning and their experience of the class. In particular the lack of interactions makes it difficult for lecturers to gauge students' understanding, which means they are not able to pick up and respond to students' misconceptions (Scott 1991).

Finding ways to increase the interactions in large classes is important, but it is not just about quantity, the quality of the interactions is also critical to learning. In traditional style lectures most of the interactions are likely to be simple questions to and from the students. More complex and therefore more valuable interactions are needed to support increased conceptual understanding (Wegerif 2013). A recent study of student–teacher interactions in a flipped active learning class found that, even though they weren't truly dialogic, they supported the development of a scientific understanding, through involving students in sense-making, guided expert thinking and wonderment questions (Wood *et al* 2018).

1.5.2 Anonymity
Related to the lack of interactions is the potential for students to feel anonymous and isolated. Students in large classes are unlikely to have one-to-one contact with a teacher and there will be few opportunities for individual feedback. This makes students feel that classes are impersonal and anonymous and that their presence is not noticed by either their peers or their teacher (Cash *et al* 2017). An in-depth case study of large classes at a UK university by Loughlin and Lindberg-Sand (2023) in health sciences found that there was a disconnect between students and the lecturer and that although there were a limited number of interactions through questions, they noted that 'most interactions are at arm's length and largely anonymous'. Loughlin and Lindberg-Sand also commented that even when questions are answered, no meaningful relationships are formed (2023). They also observed that students found such interactions difficult even when they were encouraged, with one student commenting: 'When lecturers [try to] involve people, I feel like everyone backs up because they're not used to it and they don't like it and they do feel self-conscious' and because of this they prefer to be anonymous and invisible. These experiences may ultimately lead to reduced attendance and a decrease in motivation (Cash *et al* 2017).

1.5.3 Heterogeneous background of students
A major challenge of teaching a large class is the diversity of the cohort: students are likely to come from a variety of backgrounds and educational experiences (Mulryan-Kyne 2010). Furthermore, in the first year, classes are commonly made up of a mixture of students planning to major in the subject, and those taking it as a subsidiary subject to fill in their timetable. This means that students will have very different levels of interest in the subject matter as well as different aims for the course, which will affect their approach to learning. For example, those not taking the subject as their major may be more focussed on just passing the exam, which can lead to a surface approach to learning (Entwistle *et al* 2002), while those majoring in the subject may be more motivated and more likely to invest in deep learning as they realise that later courses will build on this core knowledge.

There is also some variety in the range of entry qualifications (although all students will need to meet the minimum requirements for the course), for example in physics some, but not all students will have taken more advanced mathematics courses (e.g., Advanced Higher qualification, or Further Maths A-level), resulting in a wide range in students' prior knowledge and understanding. This variation in level of knowledge makes it difficult to strike the right balance between providing enough support to the less able students and offering an opportunity to challenge the more able students. A danger is that the teacher focuses on the 'middle of the cohort'

(Allais 2014) while students who are struggling get left behind, and those who may be considered advanced may be left disengaged (Arvanitakis 2014).

1.5.4 Assessment and feedback

Designing effective assessment and feedback approaches in large classes is a particular challenge as the associated workload can increase dramatically as student numbers increase. Approaches that work well in small classes, such as providing individualized feedback in a timely manner at multiple points through the semester, are likely to be too time-consuming to scale directly to large classes. Because of these difficulties, there is a tendency for the summative assessment in a large class to rely heavily on one final exam at the end of the course. Such a high stakes assessment does little to promote student learning, may lead to a surface approach to learning (Entwistle *et al* 2002), can cause extreme test anxiety in students (Harris *et al* 2019) and will not involve feedback that students can use to improve, since the course is already at an end. See chapter 5 for a discussion of modern approaches to assessment in large classes.

Formative assessment is a particularly important aspect of good learning design, particularly if it is combined with timely feedback to students about their learning, which they can use to improve future work (Black and Wiliam 2009, Chickering and Gamson 1987). Yet large classes, particularly those taught with a didactic approach, often contain very little formative assessment. In this approach opportunities for feedback are limited to the teacher asking 'does anyone have any questions?', which is commonly met with silence.

A further challenge for designing assessments in large classes is the heterogeneous nature of the class. As part of the assessment design, care should be given to inclusivity. Assessment activities should be designed so that all students in the class have the same opportunities to demonstrate their learning achievements. A successful inclusive assessment design will eliminate the need to adjust assessment activities after the event.

1.6 Benefits of large classes

1.6.1 Social aspects

Arguments in favour of large classes point to the benefit of the collective experience of learning the same things at the same time (Collins 2014, Arvanitakis 2014). Collins for example proposes that the lecture allows for the 'creation of an intellectual environment that presents an amalgamation of key ideas and an overview of the topic' (Arvanitakis 2014). There are also social benefits of meeting, interacting with, and forming connections with others. Although students have reported feeling anonymous and isolated in large classes (Cash *et al* 2017), in-depth qualitative work on students' experiences of large classes found that even those not designed with interactivity at the centre provide beneficial social contact which enables students to both get help from their peers if they need it and to find that they are not alone in not grasping the more difficult concepts in the course (Loughlin and Lindberg-Sand 2023).

1.6.2 Diversity

Efforts towards widening participation of under-represented categories of students at university enable an increasing number of 'non-traditional' students to attend classes, for example, students with caring responsibilities. Large classes, therefore, have the potential to reflect the rich diversity of our society, under the condition that institutional or systemic barriers to entry of under-represented groups are as few as possible. It helps students from minority groups feel less isolated if they are not alone in a peer group. The diversity of voices, talents and experiences also brings more creativity and boosts problem-solving in a group. This reflects how large research collaborations work to solve societal challenges, or how universities themselves tend to combine forces to become more resilient and offer greater opportunities to students and staff.

1.7 Approaches to making large classes successful

A systematic review of research into what influences the effectiveness of large classes by Jerez *et al* found five conditions which make large classes successful: (1) student–teacher and student–student interaction, (2) implementation of active learning strategies, (3) classroom management, (4) students' motivation and commitment, and, (5) the use of online teaching resources (Jerez *et al* 2021).

As discussed above introducing high quality and diverse types of interactivity is a challenge in large classes. One approach, which will be explored in detail in later chapters is to introduce small-group activities, through pedagogies such as Think-Pair-Share (Kothiyal *et al* 2013) and Peer Instruction (Mazur 1999). This gives students the opportunity to interact with their peers, to learn from each other, to discuss the material in depth and to practice making their thinking visible (Smith *et al* 2009, Wood *et al* 2014). Beyond the clear learning benefit, this type of approach also increases the opportunities for social interactions which reduce the feeling of isolation.

Introducing teacher–student interactions is harder, both because of the natural reticence of students to speak out in front of hundreds of their peers and because only a small number of students will ever get the chance to take an active role in the discussion—most will be passive observers of the interaction. Nevertheless, this approach has an important role to play in connecting the teacher with the student group and in creating a community of enquiry (Garrison 2009). It is also important for 'sense-making' and, as pointed out by Turpen and Finkelstein (2010) it can be particularly valuable if students are encouraged to explain their thinking for incorrect answers as well as for their correct answers.

The strategies discussed above can all be described as active learning approaches. Not only do they increase the opportunities for student–student and student–teacher interactions, they have been shown to have a range of benefits for students. These include improved performance on exams and concept inventories (Freeman *et al* 2014), reduced attrition rates (Lasry *et al* 2008) and helping to close the gap between under-represented minority students (URM) and well represented student groups (Ballen *et al* 2017).

Classroom management also plays an important role in large classes, and ultimately the success of active learning approaches depends on what is termed the 'idioculture' which can be described as 'a system of knowledge, beliefs, behaviours, and customs' (Finkelstein 2005) shared by all members of the group. When applied to a learning environment the idioculture affects the class climate, attitudes and approaches to and within a class, influencing the nature of the interactions between students and teacher. These class 'norms' can be set at the start of the course by making it clear to students what to expect in the classes (i.e., that questions are welcomed, that mistakes are learning opportunities), by modelling the behaviour that teachers want to see, and explaining to students the reasons behind a particular pedagogical approach.

Another way to support large classes is the provision of online teaching resources. These include in-class voting systems which encourage interaction such as Kahoot, Mentimeter and Socrative, out of class learning systems such as PeerWise (Bates *et al* 2011) which encourage students to create their own questions, STACK (see, e.g., Sangwin 2007) for solving mathematical problems with immediate feedback and Padlet (see, e.g., Ellis 2015) for opportunities to discuss and connect with other students. Online resources also include digital notes and lecture captures which, when used appropriately, can support students to take control of their learning (Wood *et al* 2021).

As discussed above, designing appropriate assessment and feedback opportunities is another challenge in large class teaching. The approaches already mentioned such as Think-Pair-Share and Peer Instruction provide instant feedback, both to the students about their level of (collective) understanding and to the teacher who can use that information to make decisions about class activities. In addition, large classes can benefit from the use of computer marked assessments, which can provide both formative and summative assessments in a less resource intense way, while also giving timely feedback to students (see chapter 6 for a detailed discussion of computer marked assessments).

1.8 Conclusion

The drive to increase the number of students attending higher education in recent decades, coupled with the proportional lack of investment in the number of teachers, means that large classes are likely to be a part of the higher education STEM landscape for the foreseeable future.

For large classes to be effective teachers need to be aware of both the challenges that exist and the way in which pedagogies such as active learning can offset those challenges. In this chapter we have discussed the difficulties of creating high quality student–student and student–teacher interactions in large classes, the potential for students to feel isolated and anonymous, the challenge of teaching classes with students from a mix of backgrounds, educational achievements and motivations, the difficulties of providing feedback to large numbers of students and of designing suitable assessments. We have also pointed to approaches, many of which will be fully explored in the following chapters, which can not only ameliorate these challenges but provide a high quality learning experience which encourages deep learning and critical thinking. These include pedagogies such as Peer Instruction which encourage student–student interactions and provides feedback to both the students and the teacher, and technology based strategies to provide online resources to students, to enable them to connect outside of class, to give instant feedback on their level of understanding and to provide effective assessments.

In conclusion, while this chapter has discussed many of the challenges of teaching, assessing and learning in large cohorts, we have also provided examples of ways in which large classes can not only overcome these challenges but can become high quality learning environments in their own right. Indeed as Hornsby and Osman (2014) observe, 'Large classes in and of themselves are no longer insurmountable obstacles in efforts to foster higher order cognitive skills and in the process to achieve innovation and knowledge based economies. Indeed they can offer as many opportunities as they do challenges'.

References

Ake-Little E, von der Embse N and Dawson D 2020 Does class size matter in the university setting? *Educ. Res.* **49** 595–605

Allais S 2014 A critical perspective on large class teaching: the political economy of massification and the sociology of knowledge *High. Educ.* **67** 721–34

Altbach P G 1992 Higher education, democracy, and development: implications for newly industrialized countries *Interchange* **23** 143–63

Arvanitakis J 2014 Massification and the large lecture theatre: from panic to excitement *High. Educ.* **67** 735–45

Ballen C J, Wieman C, Salehi S, Searle J B and Zamudio K R 2017 Enhancing diversity in undergraduate science: self-efficacy drives performance gains with active learning *CBE—Life Sci. Educ.* **16** ar56

Bates S, Galloway R and Mcbride K 2011 *Using PeerWise for Formative Peer eAssessment in Introductory Physics Courses Online* http://tinyurl.com/7qrxtqv

Black P and Wiliam D 2009 Developing the theory of formative assessment *Educ. Assess. Eval. Acc.* **21** 5–31

Bligh D A 1998 *What's the Use of Lectures?* (Bristol: Intellect Books)

Bloom D E, Canning D and Chan K 2006 *Higher Education and Economic Development in Africa* vol 102 (Washington, DC: World Bank)

Cash C B, Letargo J, Graether S P and Jacobs S R 2017 An analysis of the perceptions and resources of large university classes *CBE—Life Sci. Educ.* **16** ar33

Campus Review 2019 https://campusreview.com.au/2019/09/the-end-of-lectures/ (accessed 11 April 2023)

Chickering A W and Gamson Z F 1987 Seven principles for good practice in undergraduate education *AAHE Bull.* **3** 7

Collins R 2014 Interaction ritual chains and collective effervescence *Collective Emotions: Perspectives from Psychology, Philosophy, and Sociology* (Series in Affective Science) ed C von Scheve (Oxford: Oxford University Press) pp 299–311

Cuseo J 2007 The empirical case against large class size: adverse effects on the teaching, learning, and retention of first-year students *J. Fac. Dev.* **21** 5–21

Driver R, Asoko H, Leach J, Scott P and Mortimer E 1994 Constructing scientific knowledge in the classroom *Educ. Res.* **23** 5–12

Ehrenberg R G, Brewer D J, Gamoran A and Willms J D 2001 Class size and student achievement *Psychol. Sci. Public Interest* **2** 1–30

Ellis D 2015 Using Padlet to increase student engagement in lectures *Proc. European Conf. E-Learning, ECEL* vol 2013 pp 195–8

Entwistle N, McCune V and Hounsell J 2002 Approaches to studying and perceptions of university teaching-learning environments: concepts, measures and preliminary findings *Enhancing Teaching and Learning Environments in Undergraduate Courses Occasional Report No. 1*

Finkelstein N 2005 Learning physics in context: a study of student learning about electricity and magnetism *Int. J. Sci. Educ.* **27** 1187–209

Freeman S, Eddy S L, McDonough M, Smith M K, Okoroafor N, Jordt H and Wenderoth M P 2014 Active learning increases student performance in science, engineering, and mathematics *Proc. Natl. Acad. Sci.* **111** 8410–5

Garrison D R 2009 Communities of inquiry in online learning *Encyclopedia of Distance Learning* 2nd edn (Hershey, PA: IGI Global) pp 352–5

Gleason J 2012 Using technology-assisted instruction and assessment to reduce the effect of class size on student outcomes in undergraduate mathematics courses *College Teach.* **60** 87–94

The Guardian 2013 https://theguardian.com/higher-education-network/blog/2013/dec/10/in-praise-of-academic-lecture (accessed 11 April 2023)

Harris R B, Grunspan D Z, Pelch M A, Fernandes G, Ramirez G and Freeman S 2019 Can test anxiety interventions alleviate a gender gap in an undergraduate STEM course? *CBE—Life Sci. Educ.* **18** ar35

Hornsby D J and Osman R 2014 Massification in higher education: large classes and student learning *High. Educ.* **67** 711–9

Hornsby D J, Osman R and De Matos-Ala J 2013 *Large-Class Pedagogy: Interdisciplinary Perspectives for Quality Higher Education* (Stellenbosch: African Sun Media)

Jerez O, Orsini C, Ortiz C and Hasbun B 2021 Which conditions facilitate the effectiveness of large-group learning activities? A systematic review of research in higher education *Learning Res. Pract.* **7** 147–64

Jungic V, Kent D and Menz P 2006 Teaching large math classes: three instructors, one experience *Int. Elect. J. Math. Educ.* **1** 1–15

Kara E, Tonin M and Vlassopoulos M 2021 Class size effects in higher education: differences across STEM and non-STEM fields *Econ. Educ. Rev.* **82** 102104

Kinnear G, Smith S, Anderson R, Gant T, MacKay J R, Docherty P, Rhind S and Galloway R 2021 Developing the FILL+ tool to reliably classify classroom practices using lecture recordings *J. STEM Educ. Res.* 1–23

Kothiyal A, Majumdar R, Murthy S and Iyer S 2013 Effect of think-pair-share in a large CS1 class: 83% sustained engagement *Proc. 9th Annual Int. ACM Conf. on Int. Computing Education Research* pp 137–44

Lasry N, Mazur E and Watkins J 2008 Peer instruction: from Harvard to the two-year college *Am. J. Phys.* **76** 1066–9

Lawrence J E 2022 Teaching large classes in higher education: challenges and strategies *Educ. Rev.* **6** 251–62

Lenton P 2015 Determining student satisfaction: an economic analysis of the National Student Survey *Econ. Educ. Rev.* **47** 118–27

Loughlin C and Lindberg-Sand Å 2023 The use of lectures: effective pedagogy or seeds scattered on the wind? *High. Educ.* **85** 283–99

Mazur E 1999 *Peer Instruction: A User's Manual* (Upper Saddle River, NJ: AAPT: Prentice-Hall)

Mohamedbhai G 2008 *The Effects of Massification on Higher Education in Africa* (Accra: Association of African University Press)

Mulryan-Kyne C 2010 Teaching large classes at college and university level: challenges and opportunities *Teach. High. Educ.* **15** 175–85

Sangwin C J 2007 Assessing elementary algebra with STACK *Int. J. Math. Educ. Sci. Technol.* **38** 987–1002

Santiago P 2008 *Tertiary Education for the Knowledge Society: OECD Thematic Review of Tertiary Education: Synthesis Report* (Paris Cedex: Organisation for Economic Co-operation and Development, OECD) http://oecd.org/dataoecd/20/4/40345176.pdf

Scott P, Asoko H M and Driver R 1992 Teaching for conceptual change: a review of strategies *Research in Physics Learning: Theoretical Issues and Empirical Studies* ed R Duit, F Goldberg and H Niederer (Kiel: Institute of Science Education) pp 310–29

Scott P 1995 *The Meanings of Mass Higher Education* (London: McGraw-Hill Education)

Smith M K, Wood W B, Adams W K, Wieman C, Knight J K, Guild N and Su T T 2009 Why peer discussion improves student performance on in-class concept questions *Science* **323** 122–4

Stains M, Harshman J, Barker M K, Chasteen S V, Cole R, DeChenne-Peters S E, Eagan M K, Esson J M, Knight J K and Laski F A 2018 Anatomy of STEM teaching in North American universities *Science* **359** 1468–70

Turpen C and Finkelstein N D 2010 The construction of different classroom norms during peer instruction: students perceive differences *Phys. Rev. Spec. Top.-Phys. Educ. Res.* **6** 020123

Wegerif R 2013 *Dialogic: Education for the Internet Age* (Milton Park: Routledge)

Winke M P and Rawal H 2018 Teaching large, mixed-ability classes *The TESOL Encyclopedia of English Language Teaching* (Wiley) pp 1–6

Wood A K, Bailey T N, Galloway R K, Hardy J A, Sangwin C J and Docherty P J 2021 Lecture capture as an element of the digital resource landscape—a qualitative study of flipped and non-flipped classrooms *Technol. Pedagogy Educ.* **30** 443–58

Wood A K, Galloway R K, Hardy J and Sinclair C M 2014 Analyzing learning during peer instruction dialogues: a resource activation framework *Phys. Rev. Spec. Top.-Phys. Educ. Res.* **10** 020107

Wood A K, Galloway R K, Sinclair C and Hardy J 2018 Teacher–student discourse in active learning lectures: case studies from undergraduate physics *Teach. High. Educ.* 1–17

Wright M C, Bergom I and Bartholomew T 2019 Decreased class size, increased active learning? Intended and enacted teaching strategies in smaller classes *Act. Learn. High. Educ.* **20** 51–62

Chapter 2

Active learning in large classes

Anna K Wood

In this chapter I will explore the concept of active learning, its relationship to flipped learning and its use in large STEM classes. I will begin with a definition of active learning, an overview of the evidence base for its benefits and discuss the challenges of implementing active learning. I will then provide an introduction to the main learning theories which support the use of active learning and examine how three different types of activity (teacher talking, student–teacher interaction and student–student discussion) contribute to learning through a discussion of the most commonly used active learning strategy in STEM: Peer Instruction.

2.1 What is active learning?

Active learning is an approach to teaching which seeks to 'engage students in the learning process' (Prince 2004). However, it is not a defined set of pedagogies, methods or procedures and there is no single agreed definition of active learning. Active learning has been described as an approach which encourages higher order processing, where students are 'involved with the information presented, really thinking about it (analyzing, synthesizing, evaluating) rather than just passively receiving it' (King 1993, p 2) and where students become critical thinkers (Owens *et al* 2020). In physics education the term 'interactive engagement' is commonly used in place of active learning, which Hake (1998) defines as activities:

> designed at least in part to promote conceptual understanding through interactive engagement of students in heads-on (always) and hands-on (usually) activities which yield immediate feedback through discussion with peers and/or instructors....

Interactive engagement, similar to the definitions above, emphasizes that students should be cognitively engaged with their learning, however it also highlights the importance of interactions for learning. Such interactions include those between the teacher and the students (as a whole class), those between the teacher and an individual student, those between students (often in small groups), as well as how students interact directly with the material.

Active learning encompasses a range of different techniques including concept maps, work-along exercises and minute papers. Two of the most commonly used active learning approaches in large STEM classes are 'think-pair-share' (Kothiyal *et al* 2013) and Peer Instruction (Mazur 1999) which will be discussed later in the chapter as well as in chapters 3, 4, 5 and 7. Other active learning approaches less suited to large classes include experiential learning and problem based learning (covered in another book in the IOP Higher Education series (Raine 2019)).

While active learning has become increasingly popular in recent years as teachers in higher education move away from traditionally taught lectures, uptake is still low. For example Stains *et al* (2018) found that 80% of STEM classes from more than 500 faculties across America used little or no student engagement. One of the aims of this book is to provide higher education teachers with the tools they need to implement active learning approaches successfully. This chapter will explore active learning and its relationship to the flipped classroom. We will discuss the evidence for the benefits of active learning, the theories which support its use, and some of the challenges of adopting it in large classes. The final section will explore three key elements of active learning (the lecturer talking, student–student discussions and student–teacher interactions) and the ways in which they support learning.

2.2 Flipped classroom

One of the teaching approaches most commonly associated with active learning and an increasingly popular way to transform large STEM classes is the flipped classroom. Broadly speaking a flipped class is one in which students' first encounter with the material happens *before* the class, freeing up class time to be spent on more in-depth thinking about the material, such as through problem solving (Abeysekera and Dawson 2015). A central aspect of this approach is that what happens during the class time involves active learning approaches. The idea of flipping the teaching has been around at least since the middle ages (Talbert 2017). It is commonplace in the arts and humanities (where students come to a seminar having prepared through reading), however, it is only more recently that such ideas became mainstream in STEM teaching. As commonly happens with major inventions a number of different teachers stumbled onto the idea independently. The most famous is Eric Mazur, a Professor of Physics at Harvard University who was shocked to find that his students did not do particularly well on a test designed to probe conceptual understanding of Newtonian mechanics (the force concept inventory (Hestenes *et al* 1992)). Their apparent success on normal physics exams was due to memorization and computation rather than deep understanding. His solution was to flip the classroom by asking students to do readings which covered the course content before the class, and then to use class time for an active learning approach he called Peer Instruction, which will be discussed below. When using this technique students' understanding improved on a range of measures (Crouch and Mazur 2001).

However, the results for the flipped classroom in the literature are mixed (Låg and Sæle 2019, Zainuddin *et al* 2019)—while many show that a flipped classroom leads to improved learning (Akçayır and Akçayır 2018) some find no improvement compared to those which are not flipped (e.g., Love *et al* 2014, Chen *et al* 2017) and a few even find it is detrimental (e.g., Hagen and Fratta 2014). This is because the flipped classroom is not a pedagogical approach, but a general strategy for organizing teaching which can be implemented in many different ways. This also means that when the flipped classroom does lead to gains it is not clear which component(s) are responsible. In order to test whether it was the flipped nature of the course or the use of an active learning approach (which is typical of in-class teaching in a flipped class) providing the benefit, Jensen *et al* (2015) compared two classes, both taught with the same material and the same active learning philosophy, but one of which was flipped and one which wasn't. The study design aimed to vary only the role of the instructor while keeping all other variables the same. They found similar learning gains from both classes and concluded that any improvements in learning in a flipped classroom could be attributed to the active leaning, constructivist style of instruction in the classes, rather than to whether or not a flipped approach was employed. This shows that the success of the flipped classroom is likely to be a result of active learning taking place, rather than flipping per se. However, in many cases flipping the class facilitates active learning by freeing up class time through the moving of core content to outside of class time.

2.3 Active learning versus direct instruction

In the higher education research literature, active learning is often contrasted with the term 'passive learning'. However, the term is problematic when it is used in this way because it conflates two different things. Active learning is an approach/philosophy of teaching. In contrast passive learning refers to the way in which the students engage with their learning (what Chi and Wylie (2014) call an 'engagement behaviour'), as a result of a particular teaching approach. As the name suggests, passive learning implies that students are passive recipients of information. This type of learning is associated with rote learning rather than with higher order thinking.

A better term for talking about the sort of learning that isn't active learning, is 'direct instruction'. This is teaching which is typically teacher-centred, focussed on content delivery, and where the teaching is highly prescriptive (Talbert 2017), although it should be noted that the term is used differently in the school setting (see, e.g., Kozioff *et al* 2000). One of the key differences between direct instruction and active learning is the teacher's role in the classroom. As the classic essay by King (1993), entitled 'From sage on the stage to guide on the side' highlights, in direct instruction the teacher is seen as an information provider, whereas in active learning the teacher is seen more as a guide on hand to support students to create their own knowledge.

Direct instruction, like active learning, is a way of approaching teaching, rather than one particular technique, so in reality there is huge variation in the quality of the teaching in both cases. A class taught with direct instruction is likely to contain more passive learning, but it could also include time spent on questions to and from the lecturer in which the students genuinely engage with the subject matter. For this reason, equating direct instruction with passive learning is unhelpful; a more productive approach is to think about the activities that take place in the class and how they support learning, as discussed in the following section. While there is a strong evidence base (see section 2.5) for using active learning, direct instruction can still be a valuable tool for instructors to use in the right circumstances. As Talbert argues, direct instruction and active learning need to be seen as 'complementary rather than opposed or even mutually exclusive' (Talbert 2017, p 14).

2.4 Characterizing active learning approaches—ICAP

One approach which can help teachers to find strategies that are effective is the ICAP framework developed by Chi and Wylie (2014) which proposes that engagement behaviours can be categorized into four modes, interactive (I), constructive (C), active (A) and passive (P). They show empirically that students become more engaged with the learning materials as activities move from being passive to active to constructive and then to interactive. In their work they equate being 'cognitively engaged' with 'active learning' as defined above. In this way they describe passive activities as ones where new knowledge is 'encoded in an isolated or encapsulated way during learning' (Chi and Wylie 2014, p 227), which tends to lead to rote memorization. Such knowledge can be recalled if similar prompts are given. In 'active' activities students manipulate information during learning which leads to filling gaps in their understanding. In constructive activities students generate inferences and rationales, they are able to link the knowledge to other areas of knowledge so they become interconnected. Finally in 'interactive' activities new knowledge and understandings are co-created between peers discussing ideas that neither could have created individually.

The different modes of engagement do not necessarily relate directly to the instruction method, but rather to how the instruction method is implemented. For example listening to a lecture can be passive (if students only listen to the lecture), active (if students take notes or copy solution steps), constructive (if students draw concept maps or ask questions) or interactive (if students defend/argue a position in small groups). Chi and Wylie (2014) argue that thinking about the mode of engagement, and which category it falls into is a more useful approach for teachers who want to increase the amount of active learning in their classrooms.

2.5 Evidence for active learning

There is substantial evidence in the literature that active learning approaches are more effective for generating deep and long lasting learning compared to those that are non-interactive and based on direct instruction.

One of the earliest and most influential studies of active learning was conducted by Hake (1998) and involved over 6000 students from 62 courses studying Newtonian mechanics. Hake measured learning through recording how well students performed at the end of the course on a standard test called the force concept inventory (FCI), compared to how they performed at the start. He found that the gain defined as

$$g = \frac{\langle \text{Post} \rangle - \langle \text{pre} \rangle}{100 - \langle \text{pre} \rangle}$$

(where brackets denote averages, Post is the post-course score and Pre is the pre-course score) was substantially higher for courses which used interactive engagement techniques (i.e., active learning) (0.22–0.7) compared to those that didn't (0.12–0.28).

Another strong piece of evidence on the efficacy of active learning was provided by Freeman *et al* (2014) who conducted a meta-analysis of 225 studies in STEM courses that examined the effect of active learning on student outcomes. They found that active learning on average increased performance on examinations and concept inventories by 0.47 standard deviations and that students in the traditionally taught classes were 1.5 times more likely to fail than students taught in active learning classes. The results were so clear that the authors concluded:

> If the experiments analyzed here had been conducted as randomized controlled trials of medical interventions, they may have been stopped for benefit—meaning that enroling patients in the control condition might be discontinued because the treatment being tested was clearly more beneficial (Freeman *et al* 2014, p 8).

Active learning has benefits beyond attainment: Lasry *et al* (2008) showed that Peer Instruction reduces student attrition rate in difficult courses, and others have shown that it decreases failure rates (Porter *et al* 2013), and improves student attendance (Deslauriers *et al* 2011). There is also evidence that active learning can help to close the gap between underrepresented minority students (URM) and well represented student groups (Ballen *et al* 2017, Styers *et al* 2018, Theobald *et al* 2020). For example Ballen *et al* (2017) found that a transition to active learning for a large (250 students) introductory STEM class closed the gap in learning gains between non-URM and URM students and led to an increase in science self-efficacy for all students. It also increased the sense of social belonging for the non-URM students which the authors demonstrated mediated the effect on the learning gains (Ballen *et al* 2017). Similarly Burke *et al* (2020) found that active learning classes were most beneficial to Hispanic students in a general chemistry class.

2.6 Barriers to implementing active learning

Despite the strong evidence for active learning, uptake of these approaches has been limited (Børte *et al* 2020, Apkarian *et al* 2021). A number of reasons for this have been identified, including the lack of support for teachers (Børte *et al* 2020), a fear of the risk involved (Apkarian *et al* 2021, Wood *et al* 2022), perception of a lack of class time for active learning (Miller and Metz 2014), time constraints for preparing active learning (Miller and Metz 2014, Wood *et al* 2022), the additional time needed to make a substantial change in teaching approach (Schneider and Pickett 2006), a lack of recognition of teaching activities in relation to promotion (Suchman 2014, Børte *et al* 2020, Wood *et al* 2022), the layout of the classroom not being conducive to active learning (Apkarian *et al* 2021) and conflict with a faculty member's professional identity (Brownell and Tanner 2012). Another worry for teachers is that students will feel negatively about active learning and this will affect their evaluations (see below).

One key issue is the lack of professional development opportunities, which Børte *et al* (2020) suggest is key to encouraging a switch to active learning. Such courses should focus on how to design teaching with the students' learning needs at the centre. This is critical because, as discussed above, active learning is not a single magic bullet which will automatically enhance learning but rather an approach to teaching that can be implemented in many different ways. Successful strategies depend on the exact choice and implementation of the activities, as well as the course content and the lecturer and students' approaches to learning and teaching (Hodges 2020). It is also important for teachers to understand not just what to do to implement active learning strategies, but also why and how a particular approach works.

Another important factor which affects teachers' approach to teaching is their conceptions about teaching practices and their beliefs about how students learn (Ho 2000). However, as active learning requires a conceptually different perspective compared with teaching through direct instruction, a change in teachers' conceptions is needed if they are to change their teaching practices (Ho 2000). In order to encourage this change Ho (2000) has developed an approach to staff development based on theories of conceptual change, which, interestingly, are the same ideas used in science teaching to help students to change to a scientific understanding.

A second challenge to the uptake of active learning is resistance from students, which can be reflected in student surveys and end of course evaluations. Reasons for this may include unfamiliarity with the approaches being used, struggling with uncertainty when there is reduced access to authoritative information and the extra effort involved to actively construct knowledge compared to learning via traditional, teacher-centred instruction (Owens *et al* 2020). Some of the issues and how to address them are covered in the case studies in this volume.

To understand this issue in more detail Deslauriers *et al* (2011) compared students' learning and students' perceptions of learning between two cohorts of students, one which was randomly assigned to an active learning class and one to a direct instruction class with identical content. As expected the students in the active learning class learned more than those in the direct instruction class. However those in the active learning cohort had a lower perception of their learning (although still positive) compared to their peers taught through direct instruction. The authors concluded that the increased cognitive effort required for active learning leads students to feel that they are learning less. This may have a negative effect on their motivation, engagement and ability to self-regulate their own learning. This finding helps to explain the negative reactions from students which are commonly reported by teachers. The authors recommend a number of strategies to counteract this issue, such as explicitly discussing the value of increased cognitive efforts early on in the course as well as giving a test or other assessment so that students can see the progress they are making.

2.7 Theoretical basis for active learning

In order to apply active learning techniques successfully, it is helpful to understand why they are beneficial. This section will discuss some of the theories which support the use of active learning. First, two genres of learning theories (cognitive learning theory and social constructivism) which come under the heading of cognitive constructivism will be discussed. Following this, I introduce transactional distance theory which highlights the psychological and communicative factors which influence learning and how this impacts students' experiences of large classes. Finally, given the importance of interactions in active learning, some theoretical concepts around discourse and dialogue will be discussed.

2.7.1 Cognitive constructivism

Cognitive constructivism focuses on the process of thinking, emphasizing that individuals learn through building their own knowledge. This happens by connecting new ideas and experiences to existing knowledge in order to form new or enhanced understandings (Bransford 2000). Cognitive constructivism underpins the philosophy of active learning. In contrast, theories of behaviourism support (some aspects of) direct instruction. B F Skinner, a prominent advocate of behaviourism, believed that learning happens when an association is formed between a certain behaviour and the consequences of that behaviour. This theory supports the use of breaking tasks into smaller components, modelling and reinforcement activities (Magliaro *et al* 2005).

2.7.1.1 Cognitive learning theory
In the cognitivist approach to learning, the brain of an individual is considered to be the active agent in learning (Otero 2003). The focus is therefore primarily on what is going on in the head of the learner during the process of learning, although this can be influenced by interactions with the environment. Students are thought to construct knowledge based on what they already know. This idea has roots in the work of Piaget who believed that knowledge construction results from learners' physical interactions with the world around them. Piaget proposed that learning happens through assimilation, accommodation and equilibration. Assimilation happens when existing cognitive schemes adapt to incorporate new knowledge, these cognitive schemes need to then be rearranged to adapt to the new information (accommodation) and finally any conflicts need to be resolved (equilibration).

Posner later built on these concepts, particularly the idea that prior knowledge influences learning, to develop conceptual change theory (Posner *et al* 1982) which has been influential in research on how students learn science. In conceptual change theory, students' prior conceptions are believed to govern how they interact with new ideas (Posner and Strike 1992). Such prior conceptions, also called naive conceptions or misconceptions, are based on how students experience the world. These naive conceptions may serve them well in everyday life, but are often at odds with a scientific explanation. Naive conceptions have been identified across STEM subjects (DiSessa 2001, Fisher and Moody 2002, Taber 2002, Nehm and Reilly 2007, Bozzi *et al* 2019). Although there are some conflicting ideas about the details of how to teach students to understand the world from a scientific perspective (Vosniadou 1994, DiSessa 2001), common approaches include giving students opportunities to make their thinking visible, presenting them with alternative ideas and giving them time to incorporate these into their own thinking (Driver 1989).

2.7.1.2 Social constructivism
In social learning theories, social relationships and communication play an active role in learning. This contrasts with cognitive theory where the world outside the head of the learner is considered to have only a passive influence on cognition. The most notable theorist in this area is Vygotsky who believed that learning happens through communication with others. He explicitly linked communication and thinking, particularly asserting that social interactions and language were the gateway to higher order mental processes. This is particularly relevant to many active learning techniques which aim to increase the opportunities for communication with both peers and with the teacher. Vygotsky found that learners can achieve more when working with someone who is more knowledgeable. The gap between the learning that an individual student can achieve on their own and the potential learning they can achieve when working with a more knowledgeable other (which can include a more advanced peer as well as a teacher) is called the 'zone of proximal development' (Vygotsky 1978). This concept led to the idea of scaffolding learning, where tasks given to students begin at a simple level, then increase in difficulty to help learners to master a subject.

2.7.2 Transactional distance theory

Although the theories discussed above acknowledge the influence of the environment to varying degrees, their aim is to explain how learning happens from a social and cognitive perspective. However, many other aspects of the context in which students learn have an impact on their learning. Transactional distance theory (TDT) provides a way to put the learning environment and the impact that this has on students' learning at the centre of our thinking. A key aspect of TDT is the idea that students can experience a feeling of psychological distance (such as feeling isolated or disengaged) and a communicative distance (such as misunderstandings), which are detrimental to learning. Moore called this experience 'transactional distance' Moore (1983). Transactional distance is influenced by three factors. The first factor is the structure of the course (how well defined the course is, and how much flexibility there is for students to make their own decisions about what they learn). The more structured the course is, the higher the transactional distance. The second factor is dialogue (including both student–student and student–teacher interactions). Both the quality and the quantity of the dialogue are important for reducing transactional distance. The third factor is learner autonomy (the extent to which students take responsibility for their own learning). Higher learner autonomy is associated with a lower transactional distance. Research has found that transactional distance has a negative impact on a variety of aspects of learning and teaching including perceived learning and satisfaction (Kara 2020) and learner engagement (Doo *et al* 2020). As a consequence of this, there has been a call for teachers to minimize transactional distance in order to positively affect student learning (Benton *et al* 2013).

Transactional distance theory was developed in the context of distance and online education, but, as Bender (2012, p 10) argues it applies to any learning environment including in-person teaching: 'if a teacher, whether online or on campus, can establish meaningful educational opportunities … then the transactional gap shrinks and no one feels remote from each other or from the source of learning'.

To date only a few studies have explored transactional distance in in-person learning (see, e.g., Ekwunife-Orakwue and Teng 2014, Rumble 2019, Doo *et al* 2020, Stöhr *et al* 2020), however, it is plausible to propose that TDT is a useful tool for thinking about how large STEM classes are experienced. Evidence for this comes from studies looking at how students experience large classes. For example large classes are often felt to be impersonal and anonymous, and students report that they do not have a sense of belonging (Weaver and Qi 2005, Shea *et al* 2006, Cash *et al* 2017). Cash *et al* (2017) observed 'an atmosphere where neither their peers nor their instructor noticed whether they were absent or attentive' and that students may find it challenging to interact with the teacher (both inside and outside scheduled classes). In large classes that are taught using direct instruction with very little interaction, either between students or between the teacher and students, it is therefore likely that students will experience a substantial level of transactional distance.

Active learning could reduce the transactional distance in a number of ways. Given that high quality dialogue is one of the most important elements of an active learning class, and that it is also associated with lower transactional distance, it is plausible to propose that active learning could help to reduce transactional distance in large classes. Indeed Bender suggests that 'If students are disengaged and not stimulated into being active learners, there can be a vast transactional distance, whether the students are under the teacher's nose or on the other side of the city' (Bender 2012, p 10). Further, a class which is designed with a combination of mini-lectures, problem solving and small group discussions will have many opportunities for students to interact both with each other and with the teacher. Social interaction with peers will help to reduce feelings of anonymity and increase social connections. These activities will also increase cognitive engagement with the material. Similarly, approaches which enable the lecturer to engage with the students, and to give feedback to them (whether that is through pre-lecture quizzes or through in-class questions) are likely to reduce transactional distance.

2.7.3 Discourse and dialogue

As discussed, interactions play a vital role in learning, supporting the development of higher order thinking, creating opportunities for students to articulate their thinking and the chance to engage in social interactions which can make students feel more connected. Such discussions, whether they occur between students or between students and a teacher, also form the cornerstone of many active learning techniques. In large classes there are limited ways in which dialogues can be introduced and they take careful thought and planning. However, it is worth noting that even short dialogues, such as question/answer exchanges can have a positive impact on learning. For this reason Ford and Wargo (2012) suggest it is helpful to view dialogues as components of a set of classroom activities which can work together to impact students' understanding of scientific concepts.

Dialogues can take many different forms and it is worth considering the way in which different types of dialogue impact learning. For Vygotsky, the aim of the interaction is to arrive at a single agreed meaning; however, Wegerif (2008) has argued that this type of dialogue is less valuable for learning than dialogue, where meaning arises through the different perspectives of each voice. This is called dialogic discourse (as opposed to monologic discourse).

The most common form of dialogue observed in learning settings (at least in schools, where most research on dialogues has taken place) is the initiation-response-feedback (IRF) pattern of discourse (Lemke 1990). In this discourse the teacher poses a question, the student answers and the teacher then gives feedback on that answer. However, this form of discourse has been associated with authoritative, monologic dialogues and is often viewed as an approach to learning consistent with rote learning rather than knowledge creation. Mortimer and Scott (2003), argue that both monologic and dialogic interactions are necessary in education, while others have found that discourse can be an in-between state, neither wholly 'authoritative' or wholly 'dialogic', but rather showing the characteristic structure of authoritative discourse, while including elements that give them a more dialogic orientation (Van Booven 2015). Understanding the ways in which different types of interactions support learning can help teachers to design teaching approaches to increase not just the quantity, but also the quality of dialogues in the class.

2.8 Active learning in practice—focus on Peer Instruction

This section will focus on three of the most common components of active learning techniques—the lecturer talking, student–student discussions and student–teacher interactions, and discuss how they support learning. These will be examined through the example of Peer Instruction, one of the most commonly used active learning techniques in large STEM classes. Other chapters in this book will give detailed practical guidance about how to implement Peer Instruction.

2.8.1 Peer Instruction

Peer Instruction was developed by Eric Mazur (1999) after discovering that the students in his classes did not have deep conceptual understanding of key concepts in physics. Peer Instruction (PI) is an active learning approach in which a multiple choice question is given to students which tests their conceptual understanding. A standard PI cycle has a number of components as shown in figure 2.1 (although in practice many variations on this are used): first the teacher introduces the question. The students are then invited to think about the question individually and then to vote, anonymously using electronic voting systems (also called clickers), or these days more commonly using their smart phones. If there is some disagreement in the class they are then asked to talk about the question in small groups and to vote again. Typically more students get the answer right after the small group discussion. The teacher may then conduct a whole class discussion in which students are asked to explain why they chose their answer. A final and important step of the process is for the teacher to explain why the right answer is correct and why the other answers are not correct. This approach helps to illuminate the misconceptions that students hold and to directly address them through discussion and examples.

2.8.2 The role of the teacher talking

There is clear evidence that active learning approaches have substantial benefits over direct instruction, and that classes where the teacher delivers a continuous monologue purely for the purposes of knowledge transfer are less likely to lead to deep learning. So it might seem strange to have a section of a chapter on active learning devoted to the role of the teacher talking. However, analyses of how much time is spent on different activities in large classes generally show that even when an active learning approach is used, teachers talk for approximately 50% of the time (Wood *et al* 2016, Stains *et al* 2018). This raises questions about the role of the teacher talking in active learning, how this is different from classes which are taught with

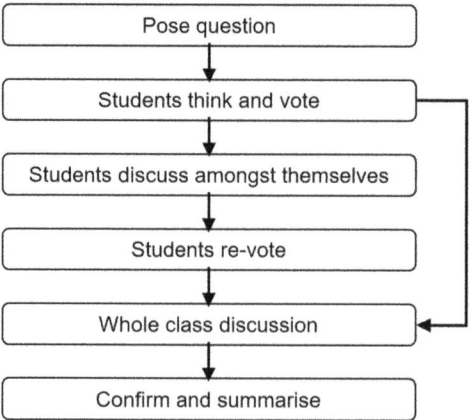

Figure 2.1. A schematic diagram showing the key steps in a Peer Instruction cycle.

direct instruction, and how it supports the other active learning activities that are part of these classes.

In direct instruction it has been suggested that uninterrupted discourse can be of benefit to students by provoking 'deep intellectual engagement' (French and Kennedy 2017, p 645), exposing students 'to the most recent developments in the field' (Biggs and Tang 2011, p 138), and enabling students to see teachers as both a researcher and a scholar (Light *et al* 2009).

While these may also be relevant to active learning classes it is notable that discussions about active learning rarely consider the specific role of the teacher talking in these classes, possibly because it is often thought of as being a 'passive' activity and considered to be of little benefit. Yet in active learning classes, the teacher talking can support students' learning in a number of different ways. If the class is flipped then students' first exposure to the content will be before the class, but the teacher may provide additional explanations in class of topics which the students found difficult. Topics which cause difficulty can be highlighted in a number of ways: by a quiz at the end of the pre-class activities, by asking for feedback during the class, or through active learning strategies such as Peer Instruction. Learning from the expertise of the teacher is an important element of opening and continuing a dialogue between teacher and students. Even during a Peer Instruction cycle there is an important role for the teacher talking. This is particularly the case at the end of the cycle where the teacher should provide an expert scientific view of how to answer the question, to ensure that all the students understand the answer. But equally importantly, they should explain why the incorrect answers are wrong, so that students can see how their thinking was flawed.

Another useful role for the teacher is in the modelling of expert thinking, explaining how to solve a problem, how to reason, how to question and even how to make (and correct) mistakes (French and Kennedy 2017). This approach can make the tacit processes of problem solving visible to learners (Dennen and Burner 2008). In an active learning environment the students should then have the opportunity to put this into practice on their own. This can be thought of as a 'cognitive apprenticeship'—where novices learn practical skills first through observation of experts and then through their own practice (Collins *et al* 1988). Although there is a place for learning by listening to the teacher solving problems, it should be noted that there is substantial evidence that listening to dialogues is more effective than listening to a monologue (Driscoll *et al* 2003, Chi *et al* 2008).

2.8.3 Student–teacher interactions

Student–teacher interactions play an important role in all large classes; they are essential for active learning classes but are also the most common form of interactions in classes taught through direct instruction. These interactions generally happen when the teacher asks a question to the students or when a student asks a question to the teacher. How often the latter happens, and the extent to which students engage with the teacher, depends on the class atmosphere, how welcome students feel to ask questions and how expected it is that they answer questions. Together these form the 'norms' of the class (Turpen and Finkelstein 2010).

It has been argued that teacher–student interactions are beneficial because a teacher can help a student to move beyond their current thinking. For example, Bishop and Verleger (2013) invoke Vygotsky's 'zone of proximal development', whereby a student can achieve more with the help of a more knowledgeable other than they can by themselves. However, interactions are varied and given the centrality of interactions in active learning, it is interesting that this topic is rarely tackled in the research literature. Understanding how these types of discourse generate and impact learning could lead to greater knowledge about implementing active learning techniques effectively.

Work by myself and colleagues (Wood *et al* 2018) looked in detail at student–teacher dialogue in a first year introductory physics class taught through a flipped active learning approach. We found three types of interaction which supported students' development of scientific ideas: (a) involving students in sense-making, (b) modelling expert thinking and (c) encouraging wonderment questions. We investigated the extent to which these interactions were dialogic and found that the pattern of discourse typically followed the common triadic IRF (initiation-response-feedback), which is considered to be authoritative, and is commonly associated with a transmissionist style rather than with active learning (Lemke 1990). However, we argue that the way in which these interactions are incorporated into the whole learning design needs to be considered. In this research the interactions were part of an active learning approach designed to help students to consider and evaluate different explanations for a scientific phenomenon. They may be considered 'ideologically dialogic' (Ford and Wargo 2012). In other words, the dialogues were part of a learner centred-approach to teaching in which interactions created an interactive learning experience and the dialogues encouraged higher order thinking, supporting deeper learning of the material.

2.8.4 Student–student interactions

A central element of active learning strategies such as Peer Instruction and think-pair-share (see chapter 3) are student–student interactions. In PI students are asked to discuss a conceptual question (often after having the chance to think about it by themselves first). While the other steps in the Peer Instruction cycle also support learning, it is arguably peer discussion which is the most important element. Peer discussion is characterized by Chi and Wylie (2014) as an interactive activity, the highest cognitive level in the ICAP framework. Such activities, at least in theory, involve both partners taking turns mutually, which the authors believe results in new knowledge and perspectives being co-created.

It is certainly true that in general, a greater percentage of students will choose the correct answer after the peer discussion, compared with after the initial stage of thinking about it alone. One possible explanation is that more knowledgeable students simply tell other students the answer. This is plausible, as it is known that students who are correct are also more confident in their answer. In this case very little would be learned by the other students. However, Tullis-Goldstone (2020) found that while confidence did have an effect on students switching answers, it wasn't the only predictor. They also found that the correctness of the answer given by a student predicted whether other students would switch to that answer, which implies that the students were convinced by the reasoning for that answer rather than the confidence of their peer. The idea that students learn something from the interactions is given weight by Smith *et al* (2009) who showed that when students were given a similar question on the same concept straight after the initial question, overall understanding was enhanced, even in groups where originally none of the students knew the correct answer.

In order to understand in more detail how these discussions lead to change, myself and colleagues (Wood *et al* 2014) recorded peer discussions that took place during PI using smart pens[1]. We found evidence that the discussions caused 'activation' of knowledge or ideas that the students already held, but which they had not thought to apply in this context. Three types of activation were observed which relate to different cognitive grain sizes. These were:

(1) Activation through knowledge elements, i.e., individual ideas or pieces of knowledge which were mentioned in the dialogue triggered a change in understanding.
(2) Activation through linkages between resources. This refers to links being made, for example, between scientific concepts and real world phenomena. An example of this is when students made a link to the TV show *The Big Bang Theory* to activate the knowledge that sin(30) is ½.
(3) Activation through control structures. This refers to the overall approach that a student is using (often tacitly), for example if they are trying to solve a problem by plugging numbers into an equation, but then something in the dialogue triggers them to change their approach to thinking about the problem through physical reasoning, or by drawing a diagram.

2.9 Conclusions

Active learning strategies transform large classes from being an event focussed on information transfer, to one where knowledge is actively constructed and students are engaged in their learning. There is significant evidence that active learning approaches lead to learning gains, reduce attainment gaps and aid retention. However, there is also a gap in uptake driven by a range of factors including time pressures, lack of recognition for teaching achievements, inadequate professional development opportunities and the potential negative attitude of students. This chapter gives an overview of the key concepts needed for teachers to begin to explore using active learning in large STEM classes. It has covered the definitions of active learning, direction instruction and passive learning and discussed how these relate to the flipped classroom. The central theoretical ideas which can support active learning have been explored and detailed examples of key interactions in active

---

[1] A smart pen is a pen which records both audio and writing electronically.

learning, student–student, student–teacher as well as the role of the teaching talking have been covered. The following chapters in this book will provide practical guidance, examples and advice for how to apply these ideas in order to implement active learning in practice.

References

Abeysekera L and Dawson P 2015 Motivation and cognitive load in the flipped classroom: definition, rationale and a call for research *High. Educ. Res. Dev.* **34** 1–14

Akçayır G and Akçayır M 2018 The flipped classroom: a review of its advantages and challenges *Comput. Educ.* **126** 334–45

Apkarian N, Henderson C, Stains M, Raker J, Johnson E and Dancy M 2021 What really impacts the use of active learning in undergraduate STEM education? Results from a national survey of chemistry, mathematics, and physics instructors *PLoS One* **16** e0247544

Ballen C J, Wieman C, Salehi S, Searle J B and Zamudio K R 2017 Enhancing diversity in undergraduate science: self-efficacy drives performance gains with active learning *CBE—Life Sci. Educ.* **16** ar56

Bender T 2012 *Discussion-Based Online Teaching to Enhance Student Learning: Theory, Practice and Assessment* (Sterling, VA: Stylus Publishing LLC)

Benton S L, Li D, Gross A, Pallett W H and Webster R J 2013 Transactional distance and student ratings in online college courses *Am. J. Distance Educ.* **27** 207–17

Biggs J and Tang C 2011 *Teaching for Quality Learning at University* (London: McGraw-Hill Education)

Bishop J L and Verleger M A 2013 The flipped classroom: a survey of the research *ASEE National Conf. Proc. (Atlanta, GA)* http://www.studiesuccesho.nl/wp-content/uploads/2014/04/flipped-classroom-artikel.pdf

Børte K, Nesje K and Lillejord S 2020 Barriers to student active learning in higher education *Teach. High. Educ.* **28** 1–19

Bozzi M, Ghislandi P, Kazuhiko T, Mami M, Motoi W, Naoto N, Pnev A B, Zhirnov A A, Gwenaëlle G and Zani M 2019 Highlight misconceptions in physics: a TIME project *Int. Technology, Education and Development Conf. 'INTED' (Valencia)* pp 2520–5

Bransford J 2000 *How People Learn: Brain, Mind, Experience, and School* (Washington, DC: National Academies Press)

Brownell S E and Tanner K D 2012 Barriers to faculty pedagogical change: lack of training, time, incentives, and… tensions with professional identity? *CBE—Life Sci. Educ* **11** 339–46

Burke C, Luu R, Lai A, Hsiao V, Cheung E, Tamashiro D and Ashcroft J 2020 Making STEM equitable: an active learning approach to closing the achievement gap *Int. J. Act. Learn.* **5** 71–85

Cash C B, Letargo J, Graether S P and Jacobs S R 2017 An analysis of the perceptions and resources of large university classes *CBE—Life Sci. Educ.* **16** ar33

Chen F, Lui A M and Martinelli S M 2017 A systematic review of the effectiveness of flipped classrooms in medical education *Med. Educ.* **51** 585–97

Chi M T, Roy M and Hausmann R G 2008 Observing tutorial dialogues collaboratively: insights about human tutoring effectiveness from vicarious learning *Cogn. Sci.* **32** 301–41

Chi M T and Wylie R 2014 The ICAP framework: linking cognitive engagement to active learning outcomes *Educ. Psychol.* **49** 219–43

Collins A, Brown J S and Newman S E 1988 Cognitive apprenticeship *Think. J. Philos. Child* **8** 2–10

Crouch C H and Mazur E 2001 Peer instruction: ten years of experience and results *Am. J. Phys.* **69** 970

Dennen V P and Burner K J 2008 The cognitive apprenticeship model in educational practice *Handb. Res. Educ. Commun. Technol.* **3** 425–39

Deslauriers L, Schelew E and Wieman C 2011 Improved learning in a large-enrollment physics class *Science* **332** 862

DiSessa A A 2001 *Changing Minds: Computers, Learning, and Literacy* (Cambridge, MA: MIT Press)

Doo M Y, Bonk C J, Shin C H and Woo B-D 2020 Structural relationships among self-regulation, transactional distance, and learning engagement in a large university class using flipped learning *Asia Pac. J. Educ.* **41** 1–17

Driscoll D M, Craig S D, Gholson B, Ventura M, Hu X and Graesser A C 2003 Vicarious learning: effects of overhearing dialog and monologue-like discourse in a virtual tutoring session *J. Educ. Comput. Res.* **29** 431–50

Driver R 1989 Students' conceptions and the learning of science *Int. J. Sci. Educ.* **11** 481–90

Ekwunife-Orakwue K C and Teng T-L 2014 The impact of transactional distance dialogic interactions on student learning outcomes in online and blended environments *Comput. Educ.* **78** 414–27

Fisher K M and Moody D E 2002 Student misconceptions in biology *Mapping Biology Knowledge* (Berlin: Springer) pp 55–75

Ford M J and Wargo B M 2012 Dialogic framing of scientific content for conceptual and epistemic understanding *Sci. Educ.* **96** 369–91

Freeman S, Eddy S L, McDonough M, Smith M K, Okoroafor N, Jordt H and Wenderoth M P 2014 Active learning increases student performance in science, engineering, and mathematics *Proc. Natl. Acad. Sci.* **111** 8410–5

French S and Kennedy G 2017 Reassessing the value of university lectures *Teach. High. Educ.* **22** 639–54

Hagen E J and Fratta D 2014 Hybrid learning in geological engineering: why, how, and to what end? Preliminary results *Geo-Congress 2014: Geo-Characterization and Modeling for Sustainability* pp 3920–9

Hake R R 1998 Interactive-engagement versus traditional methods: a six-thousand-student survey of mechanics test data for introductory physics courses *Am. J. Phys.* **66** 64–74

Hestenes D, Wells M and Swackhamer G 1992 Force concept inventory *Phys. Teach.* **30** 141–58

Ho A S 2000 A conceptual change approach to staff development: a model for programme design *Int. J. Acad. Dev.* **5** 30–41

Hodges L C 2020 Student engagement in active learning classes *Active Learning in College Science* (Berlin: Springer) pp 27–41

Jensen J L, Kummer T A and Godoy P D d M 2015 Improvements from a flipped classroom may simply be the fruits of active learning *CBE—Life Sci. Educ.* **14** ar5

Kara M 2020 Transactional distance and learner outcomes in an online EFL context *Open Learn. J. Open Distance E-Learn.* **36** 45–60

King A 1993 From sage on the stage to guide on the side *Coll. Teach.* **41** 30–5

Kothiyal A, Majumdar R, Murthy S and Iyer S 2013 Effect of think-pair-share in a large CS1 class: 83% sustained engagement *Proc. 9th Annual Int. ACM Conf. on Int. Computing Education Research* pp 137–44

Kozioff M A, LaNunziata L, Cowardin J and Bessellieu F B 2000 Direct instruction: its contributions to high school achievement *High Sch. J.* **84** 54–71

Låg T and Sæle R G 2019 Does the flipped classroom improve student learning and satisfaction? A systematic review and meta-analysis *AERA Open* **5** 2332858419870489

Lasry N, Mazur E and Watkins J 2008 Peer instruction: from Harvard to the two-year college *Am. J. Phys.* **76** 1066–9

Lemke J L 1990 *Talking Science: Language, Learning, and Values* vol 1 (New York: Ablex Publishing)

Light G, Calkins S and Cox R 2009 *Learning and Teaching in Higher Education: The Reflective Professional* (Thousand Oaks, CA: SAGE)

Love B, Hodge A, Grandgenett N and Swift A W 2014 Student learning and perceptions in a flipped linear algebra course *Int. J. Math. Educ. Sci. Technol.* **45** 317–24

Magliaro S G, Lockee B B and Burton J K 2005 Direct instruction revisited: a key model for instructional technology *Educ. Technol. Res. Dev.* **53** 41–55

Mazur E 1999 *Peer Instruction: A User's Manual* (Upper Saddle River, NJ: AAPT: Prentice-Hall)

Miller C J and Metz M J 2014 A comparison of professional-level faculty and student perceptions of active learning: its current use, effectiveness, and barriers *Adv. Physiol. Educ.* **38** 246–52

Moore M G 1983 The individual adult learner *Educ. Adults Adult Learn. Educ.* **1** 153–68

Mortimer E and Scott P 2003 *Meaning Making in Secondary Science Classrooms* (New York: McGraw-Hill International)

Nehm R H and Reilly L 2007 Biology majors' knowledge and misconceptions of natural selection *Bioscience* **57** 263–72

Otero V K 2003 Cognitive processes and the learning of physics, part I: the evolution of knowledge from a Vygotskian perspective *Proc. of the Int. School of Physics 'Enrico Fermi' Course CLVI Part of the Research on Physics Education Series (15–25 July 2003)* pp 409–46 http://www.citeulike.org/group/10888/article/9775536

Owens D C, Sadler T D, Barlow A T and Smith-Walters C 2020 Student motivation from and resistance to active learning rooted in essential science practices *Res. Sci. Educ.* **50** 253–77

Porter L, Bailey Lee C and Simon B 2013 Halving fail rates using peer instruction: a study of four computer science courses *Proc. 44th ACM Technical Symp. on Computer Science Education* pp 177–82

Posner G J and Strike K A 1992 A revisionist theory of conceptual change *Philos. Sci. Cogn. Psychol. Educ. Theory Pract.* **147**

Posner G J, Strike K A, Hewson P W and Gertzog W A 1982 Toward a theory of conceptual change *Sci. Educ.* **66** 211–27

Prince M 2004 Does active learning work? A review of the research *J. Eng. Educ* **93** 223–31

Raine D J 2019 *Problem-Based Approaches to Physics* (Bristol: IOP Publishing)

Rumble G 2019 *The Planning and Management of Distance Education* (Milton Park: Routledge)

Schneider R and Pickett M 2006 Bridging engineering and science teaching: a collaborative effort to design instruction for college students *Sch. Sci. Math.* **106** 259–66

Shea P, Li C S and Pickett A 2006 A study of teaching presence and student sense of learning community in fully online and web-enhanced college courses *Internet High. Educ.* **9** 175–90

Smith M K, Wood W B, Adams W K, Wieman C, Knight J K, Guild N and Su T T 2009 Why peer discussion improves student performance on in-class concept questions *Science* **323** 122–4

Stains M, Harshman J, Barker M K, Chasteen S V, Cole R, DeChenne-Peters S E, Eagan M K, Esson J M, Knight J K and Laski F A 2018 Anatomy of STEM teaching in North American universities *Science* **359** 1468–70

Stöhr C, Demazière C and Adawi T 2020 The polarizing effect of the online flipped classroom *Comput. Educ.* **147** 103789

Styers M L, Van Zandt P A and Hayden K L 2018 Active learning in flipped life science courses promotes development of critical thinking skills *CBE—Life Sci. Educ.* **17** ar39

Suchman E L 2014 Changing academic culture to improve undergraduate STEM education *Trends Microbiol* **22** 657–9

Taber K 2002 *Chemical Misconceptions: Prevention, Diagnosis and Cure* vol 1 (London: Royal Society of Chemistry)

Talbert R 2017 *Flipped Learning: A Guide for Higher Education Faculty* (Sterling, VA: Stylus Publishing, LLC)

Theobald E J, Hill M J, Tran E, Agrawal S, Arroyo E N, Behling S, Chambwe N, Cintrón D L, Cooper J D and Dunster G 2020 Active learning narrows achievement gaps for under-represented students in undergraduate science, technology, engineering, and math *Proc. Natl. Acad. Sci.* **117** 6476–83

Tullis J G and Goldstone R L 2020 Why does peer instruction benefit student learning? *Cogn. Res. Princ. Implic.* **5** 1–12

Turpen C and Finkelstein N D 2010 The construction of different classroom norms during peer instruction: students perceive differences *Phys. Rev. Spec. Top.-Phys. Educ. Res.* **6** 020123

Van Booven B C D 2015 Revisiting the authoritative–dialogic tension in inquiry-based elementary science teacher questioning *Int. J. Sci. Educ.* **37** 1182–201

Vosniadou S 1994 Capturing and modeling the process of conceptual change *Learn. Instr.* **4** 45–69

Vygotsky L S 1978 Interaction between learning and development *Mind and Society: The Development of Higher Psychological Processes* ed M Cole, V John-Steiner, S Scribner and E Souberman (Cambridge, MA: Harvard University Press) pp 79–91

Weaver R R and Qi J 2005 Classroom organization and participation: college students' perceptions *J. High. Educ.* **76** 570–601

Wegerif R 2008 Dialogic or dialectic? The significance of ontological assumptions in research on educational dialogue *Br. Educ. Res. J.* **34** 347–61

Wood A K, Christie H, MacKay J R and Kinnear G 2022 Using data about classroom practices to stimulate significant conversations and aid reflection *Int. J. Acad. Dev.* https://www.tandfonline.com/doi/full/10.1080/1360144X.2022.2103817

Wood A K, Galloway R K, Donnelly R and Hardy J 2016 Characterizing interactive engagement activities in a flipped introductory physics class *Phys. Rev. Phys. Educ. Res.* **12** 010140

Wood A K, Galloway R K, Hardy J and Sinclair C M 2014 Analyzing learning during peer instruction dialogues: a resource activation framework *Phys. Rev. Spec. Top.-Phys. Educ. Res.* **10** 020107

Wood A K, Galloway R K, Sinclair C and Hardy J 2018 Teacher–student discourse in active learning lectures: case studies from undergraduate physics *Teach. High. Educ.* **23** 818–34

Zainuddin Z, Haruna H, Li X, Zhang Y and Chu S K W 2019 A systematic review of flipped classroom empirical evidence from different fields: what are the gaps and future trends? *Horizons* **27** 72–86

IOP Publishing

Effective Teaching in Large STEM Classes

Anna K Wood

Chapter 3

Practical approaches to active learning

Alison Voice

In this chapter I look at the practical details of implementing active learning in large classes. I begin by exploring the value of discussion between students. I elucidate the use of conceptual questions with peer discussion to shift students' focus from wanting to know the right answer, to developing a deeper understanding of the subject. I explore a variety of question styles and give practical suggestions for how to source or create such questions relevant to the material being taught. Finally I discuss how different room layouts, and students' learning preferences must be considered in order to provide an inclusive environment where all students feel able to engage.

3.1 Features of active learning in the classroom

We have seen in chapter 2 that active learning is a technique that aims to engage students in their learning (Prince 2004) rather than merely receiving it passively. The aim of active learning is to develop deep thinking and processing of new information (King 1993, p 2). Underpinning active learning is the theory of constructivism, (Piaget and Inhelder 1969) which holds that students learn when they grapple with ideas to build their own picture of the discipline. Such an approach has been shown to have a deeper and longer lasting effect than direct instruction (Freeman *et al* 2014). However, direct instruction can still have a place, particularly in laying a foundation of knowledge: as discussed by (Talbert 2017, p 14) it is the combination of active learning, building on direct instruction, that produces a powerful learning experience.

Active learning encourages students to ponder on new material, to strive for deep understanding, to see links between topics, to consider implications of the theory and to appreciate how this can be applied in different situations. It can help to shift students' focus from just wanting to know the right answer, to wanting to understand the subject. This difference between the former 'surface' approach and the latter 'deep' approach to learning is discussed by Entwistle (1991) and explored in the context of active learning by Wilson and Fowler (2005). These authors have demonstrated that students who had previously been surface learners adopted a deeper learning approach when exposed to active learning techniques. This was attributed to the greater expectations and responsibilities of the learner in such activities. However, the nature of the final assessment can also influence the extent to which students feel deep learning is valued, and hence affect their level of engagement with it (Thomas and Bain 1984).

A key component of an active learning approach is creating opportunities for interactions, whether this is interactions between students or between students and the teacher. One of the most effective ways to introduce interactions into the classroom is through questions, particularly questions which engage the students in deep thinking and processing of new ideas (Chin 2007, Makhene 2019). Carefully designed questions can also surface (and address) students' misconceptions which may hinder understanding and distort new learning. As advocated by Novak (2002) an important goal of STEM teaching is to help students to develop correct conceptions. Suggested types of questioning and ways to deploy these in large classes are explored in sections below.

The ideas of Vygotsky (1978) are also helpful here. He postulated that learning happens through communication with others, where it is the use of language that facilitates the higher mental processes needed to make sense of new ideas. In particular he emphasized the value gained through such conversations by defining a 'zone of proximal development' as the improvement in learning when discussing with a more knowledgeable person over the learning achieved by a student when thinking alone.

In a large class the most readily available 'others' are students' own peers, and peer discussion is thus a common feature of active learning classes in a wide variety of scientific disciplines, e.g., physics (Mazur 1997), biology (Smith *et al* 2009) and computer science (Porter *et al* 2011). This means that students do not have to know or recall everything themselves, they can pool ideas and relevant information. Suggestions from one person will trigger thoughts in another, and in this way most students can feel able to contribute and gain from the discussion. In fact Porter *et al* (2011) reports that even if the group discussion leads to the wrong answer students still have the view that 'the discussion itself resulted in a better understanding of the relevant concepts'.

As expounded by Kuhn and Udell (2003) one of the main benefits of peer discussion is in negotiating ideas and confronting alternative explanations. There are differing views on whether students need to reach a consensus in their discussion (Lasry *et al* 2016) but encouraging students to explain and justify their answers to others requires them to compile a logical argument for their view. It is this back and forth group discussion that is a powerful way for students to hone their understanding and become deep learners. In addition, requiring students to select an answer (using a voting system) encourages them to take accountability for their answer.

The sections below give practical suggestions for developing effective questions, organizing a large class to engage in active learning with peer discussion and accommodating students with different learning preferences, in particular neurodiversity.

3.2 Question types

This section presents different styles of questions, and discusses how they develop different types of thinking and learning.

Bloom (1956) developed a taxonomy of learning, categorized as a hierarchy of activities. These have subsequently been updated by Anderson and Krathwohl (2001) and are shown in figure 3.1. The most fundamental of these is 'remember', and this underpins all higher categories as students need to recall prior learning in order to work with it. Whilst it is good for students to be tested, or test themselves, on how well they can remember the material (laws, equations, facts, diagrams, processes, etc) this can often be done out of class and as an individual exercise. A better use of class time, with its opportunity for peer interaction, is in the higher skills identified by Bloom, namely understand, apply, analyse and evaluate. These activities require students to identify, or remember, the fundamental knowledge of the subject, but not to stop there. Students need to work with this knowledge to understand cause and effect, to predict behaviour, to interpret diagrams, to apply concepts to diverse situations and to make judgements.

The highest level identified by Bloom, 'create', which by its very nature requires open-ended output, is perhaps harder to incorporate and evaluate within a large class session. Such creations could include a diagram, set of instructions, computer code, etc. However such activities can be accommodated if students work together to create an output, and peers then critique this output, helping students to refine their work. In this way underlying themes and alternative solutions can be shared for all

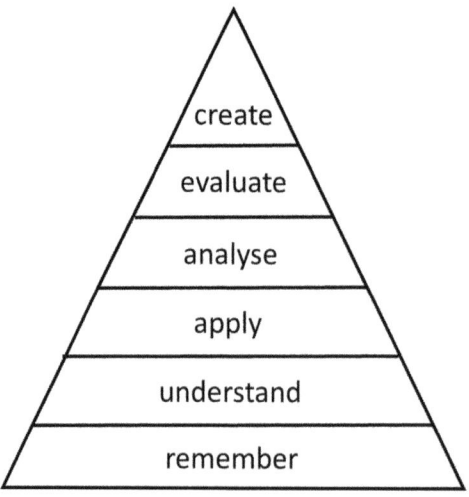

Figure 3.1. Bloom's taxonomy of learning. Levels as described. After revision by Anderson *et al* (2001).

to reflect on. For literature on the effectiveness of peers giving feedback at this level of Bloom's taxonomy see Rodgers *et al* (2012).

This raises the issue that questions can be closed (with a right answer) or open (with many and varied applicable responses). For much of the active learning and development of understanding in STEM being discussed in this book, closed questions are most effective, although this does not mean that the questions need to have a short answer. Closed questions can have suggested answers from which to choose (multiple choice questions, referred to here as MCQs) or have free text, number or diagram as a response, to be judged against the 'correct' answer. In a large class setting MCQs has obvious advantages. They give students a defined way of expressing their answer, and this gives immediate feedback to both the students and the teacher on whether good understanding has been achieved, and by what percentage of the class. But free text can be more powerful in determining the nuances in level of understanding, and eliminates the possibility that students just guess the answer. For further exploration of these ideas see discussion by Jordan in chapter 6, or see other literature (Lin and Singh 2013, Shaibar and van der Vleuten 2013, Simon and Snowdon 2014).

3.2.1 Designing questions

When writing active learning questions there is no need to 'reinvent the wheel'. Many people have written excellent questions and such questions can be accessed in text books or online in question banks or other web sites. Searching the internet using the phrase 'clicker questions in...' and naming your topic, can usually identify many questions which can be used or adapted to suit the class. The word 'clicker' originates from the handsets originally used to vote on answers to active learning MCQs.

When adapting existing questions or writing from scratch it is best to focus on the learning outcomes for the particular topic being taught (i.e., what should students be able to do with the material) and the misconceptions typically held by students in that area. Incorporating these common misconceptions as options alongside the correct answer not only affords identification of what the students in the class don't understand, but it provides detail of the way in which their thinking is misguided. One method of teachers identifying common misconceptions in advance is from mistakes made by students in coursework or exams. These can be mistakes in calculations, explanations, procedures, diagrams, or interpreting graphs etc. For discussion of a few examples see the blog by Michael Farabaugh (2020).

Table 3.1 presents a variety of question types, with comments on their suitability to promote active learning and address the different aspects of Bloom's taxonomy. Examples of these question types are discussed below. There is also much published literature about writing such questions, for example Johnstone and Ambusaidi (2000), Nash and Krauss (2015), Berk (1996) and useful hints from the University of IOWA. In chapter 6 of this book Jordan provides further insights for honing MCQ questions.

Table 3.1. Example of question types showing their suitability for active learning and the categories of Bloom's taxonomy tested.

Question type	Advantages or disadvantages	Bloom's categories	Does it promote active learning?
Recall	Students often just have to identify the answer which matches with a fact they have learned. No real thinking or understanding is required.	Remember	No. Whilst the student needs to be proactive to undertake the testing, it just involves recall of what has been taught.
Routine procedure	Students carry out a routine procedure or calculation that could be learned. The numbers in such a calculation can obscure the underlying relationship.	Remember Apply	No. Whilst the student needs to be proactive to undertake the practice, the level of thinking is not probing.
Relationship	Students need to recall an equation or relationship, and determine which behaviour will happen.	Remember Understand Apply Analyse	Yes. Predicting behaviour involves students in thinking about trends and links in the material, which goes beyond what has been taught.
Interpretation	This involves students understanding a diagram, graph or table, and reading off information, or making a judgement, based on other recalled or understood material.	Remember Understand Apply Analyse Evaluate	Yes. This gets to the heart of active learning, challenging students to see connections and meaning in the material.
Synthesis	Students need to combine several items of knowledge to interpret a situation.	Remember Understand Apply Analyse Evaluate	Yes. This requires students to examine the material in both breadth and depth.
Multiple response	This probes wider understanding as students cannot simply find one correct answer. They have to test out all scenarios.	Remember Understand Apply Analyse	Yes. This challenges students to think deeply to build, and test, their own mental model of the subject.

To highlight the features of questions that promote active learning in large classes, we start with one that is not really suitable, figure 3.2. This is a recall question that purely asks students to remember the symbol for a chemical element. Such a question is valuable in its own right to help students know what is important to learn, and to check if they have learned it. Remembering is also an important part of active learning, for students to be able to select the relevant material to bring to

> For which chemical element is the symbol Sn used?
>
> A Silicon
> B Silver
> C Tin
> D Tungsten

Figure 3.2. A recall question. Students either know this or not, and this question type does not promote active learning (ANS = C).

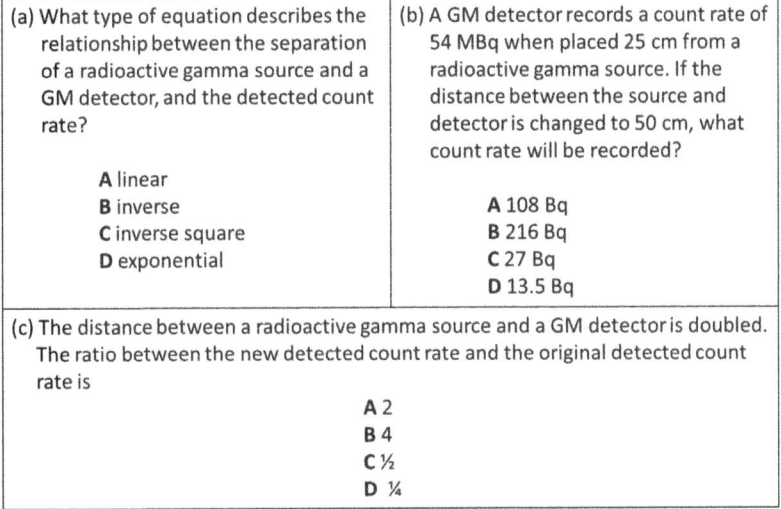

Figure 3.3. Three ways to ask students about the same concept. Panel (a) is a recall question (ANS = C). Panel (b) is a routine procedure (ANS = D). Panel (c) is a relationship question (ANS = D) and promotes good active learning to reveal and consolidate the concept.

peer discussion, but of itself this question does not require deep thinking or discussion and thus can best be undertaken individually outside class time.

Figure 3.3 illustrates three ways to ask students about the same concept. Question (a) is essentially another example of a recall question, since this topic is taught as the 'inverse square law for gamma radiation' and students could answer the question with no knowledge of the mathematical relationship between count rate and distance. Question (b) hides the relationship with too many numbers. Solving this is a routine procedure that students should be able to perform, but it does not really promote active learning, as there is not much to discuss and no particular insight to be gained. Question (c) is a relationship question and involves numbers only as a

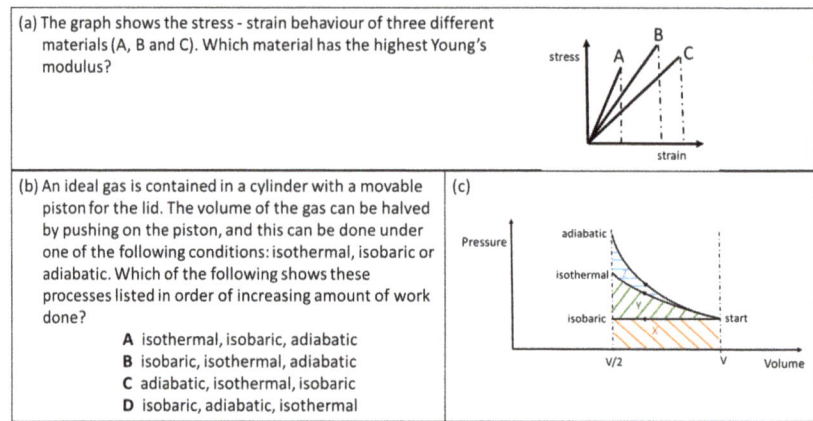

Figure 3.4. (a) An interpretation type question, promoting active learning to understand, analyse and evaluate the diagram and its meaning (ANS = A). (b) A synthesis type question, where students should be able to use their knowledge and understanding to sketch the graph shown in (c) to be able to evaluate the answer (ANS = B).

means of expressing the desired relationship (that as distance between source and detector doubles the measured count rate falls by a factor of 4). This embodies the understanding of an inverse square law and requires students to think about the implications of it, and as such this type of question is an excellent way to promote active learning.

An example of an interpretation type of question is shown in figure 3.4(a), where students are required to interpret what the diagram is showing. They need to remember that Young's modulus is given by the ratio of stress/strain and understand that this is represented by the slope of the lines on the graph. They need to apply that knowledge to the given diagram and evaluate which one has the highest slope (line A) whilst not being distracted or confused by the fact that line B extends the furthest on the stress axis and line C extends the furthest on the strain axis. This question promotes excellent discussion, and hence active learning, as many concepts are embodied in one diagram, and students need to disentangle them to answer the question posed. In fact the same diagram could also be used to ask which material is the strongest (B) and which is the most extensible (C).

Going a stage further figure 3.4(b) shows a synthesis type question where students' understanding is clarified through constructing their own diagram to reveal a pattern or relationship. In this question students need to remember that 'work done' is given by the area under the curve on a graph of pressure versus volume, and understand what that means. They also need to remember the shape of the curve for each of the given processes (isothermal, isobaric and adiabatic). By applying the specific information given in the question they should be able to sketch curves (akin to that shown in figure 3.4(c)) from which they can analyse the areas and rank them in order. (This reveals the isobaric process to require the least work to be done (area X), followed by isothermal (area $X+Y$), followed by adiabatic (area $X+Y+Z$).)

An object is moving at constant speed in a horizontal circle of constant radius. Which of the following properties of the object are also constant during the motion? 1 Linear momentum 2 Acceleration 3 Angular velocity Response Key:				
A	B	C	D	E
1 only	3 only	1 and 2	2 and 3	1, 2 and 3

Figure 3.5. A multiple response question which can promote good active learning and peer discussion. (ANS = B. All quantities have constant magnitude, but only angular velocity has constant magnitude and direction.)

Questions similar to those in figure 3.4 promote excellent active learning and peer discussion. They involve many different aspects such as interpreting the question, assembling the required knowledge, combining information to analyse and determine the answer. These are complex activities and thus prompt discussion, and encourage students to learn from listening and trying to explain the concepts involved.

Having more than one correct response (multiple response type, see figure 3.5) encourages students to probe a deep level of understanding, as they can no longer purely pick out an answer they know to be correct, but need to make a decision on each statement. Obviously the response key does not allow all combinations of true and false, but a cleverly worded question can force students to focus on the important points to be understood.

The questions discussed in figures 3.2–3.5 merely serve as examples to illustrate and provide food for thought. Beyond MCQs, which have one or multiple correct options, questions can also be posed as individual statements where students need to decide if each is true or false, or questions can be written where students need to determine the correct explanation for something, or where students need to rearrange options in a correct order. The exact nature of the question is not important, what matters is how well the question addresses the concepts and content of the discipline and the cognitive thought processes that can be elucidated by active learning.

3.3 Methods of deploying active learning in large classes

There is no one right method of deploying active learning in STEM classes, and teachers can design their activities to suit the subject material, the students and the practicalities of time and location. Three different versions of incorporating active learning are presented below to showcase a range of options. Firstly a fully flipped approach is described where all content is delivered outside class and sessions with students and teacher are all dedicated to active learning. Secondly an integrated approach is described where all sessions combine direct teaching of new content with active learning. A third system is presented where different sessions alternate between direct teaching and active learning.

Regardless of the exact interplay between content delivery and active learning, the methods of engaging students in active learning remain much the same. The most common and widely researched approaches are peer instruction (Crouch and Mazur 2001) as discussed in chapter 2, and think-pair-share (Lyman 1981, Kothiyal *et al* 2013). Both of these methods involve students thinking about conceptual questions individually, and then in discussion with their peers. Many subtle variations exist and Vickrey *et al* (2015) discuss these in a review article.

The main elements of these approaches have already been summarized in figure 2.1. Voting is usually via a mobile app, where the teacher can see the students' responses before revealing to the class. If about 80% or more of the class vote correctly in the individual vote, it is common for the teacher to acknowledge the correct answer, give a quick explanation of the correct and incorrect options, and move on, omitting the peer discussion.

There are variations where students are not asked to vote individually before the peer discussion, but research has shown that committing to an answer enhances students' engagement (Crouch *et al* 2004). There are also variations where the peer group are asked to reach a consensus, rather than vote individually, and this has been shown by Lasry *et al* (2016) to help students appreciate that their peers do not have a vastly advanced level of understanding from them, thus building their confidence to persevere.

3.3.1 Flipped classroom

In the flipped classroom method (discussed in section 2.2 and in more detail in the case-studies, particularly chapters 7–10) the content is presented outside class, and all of the class sessions are devoted entirely to active learning, to develop understanding, and explore implications and applications of the material. This allows the class time to be student-centred and encourages students to assume more responsibility for their learning. To promote the best classroom interaction it helps to support students in studying the out-of-class material so they can be prepared in advance. Providing the material in manageable chunks, with engaging media (e.g., videos) and setting up discussion boards for students to ask questions if needed can all facilitate study by students. Many teachers use short quizzes before each class as an incentive for students to keep up to date, and to reveal which aspects of the material students find most difficult. This can then inform and help the teacher to plan their active learning sessions. See Chen *et al* (2014) and Cummins *et al* (2016) for further ideas and discussion of the flipped classroom method. However, flipped classes can also have disadvantages for students who find it hard to regulate their study causing them to get behind with the material and thus be unable to participate or benefit from the class discussions. For a detailed review of the benefits and challenges of flipped classrooms see Akçayır and Akçayır (2018).

3.3.2 Integrated teaching and active learning

In this method content is delivered in class with frequent interspersion of active learning questions and discussion. This allows teachers to adapt to students' responses in real time. It also keeps all the delivery within class time thus eliminating the issue of whether or not students arrive prepared for class. However, this can result in less material being covered overall, unless contact hours are increased. This should not matter, because what is lost in breadth can be more than compensated for in depth of understanding. Typically each delivery section lasts about 7–10 min and is followed by one active learning question, tackled individually and via peer discussion in the manner set out in figure 2.1. In this scenario questions should not be too complex in design so misconceptions can be revealed easily, to aid the teacher when explaining the answers and delivering the next 7–10 min section.

3.3.3 Active learning workshops

This method allows for the same kind of in-depth active learning sessions as the flipped classroom, but in this case the content is delivered in a previous in-class session, rather than out of class. This gives time for concentrated active learning and peer discussion, whilst reducing the risk of students arriving unprepared. A full session devoted to active learning allows the method in figure 3.6(a) to be employed. This type of session works well in a flat-floored room where students can sit in groups around tables, but if this type of venue is not available then it can be facilitated with students discussing with their immediate neighbours in a lecture theatre. See section 3.4 for more practical details of room layout and ideas for running these sessions.

About 5–8 questions can be released at once, giving students time to explore one or more concepts in detail. The silent individual working gives all students time to

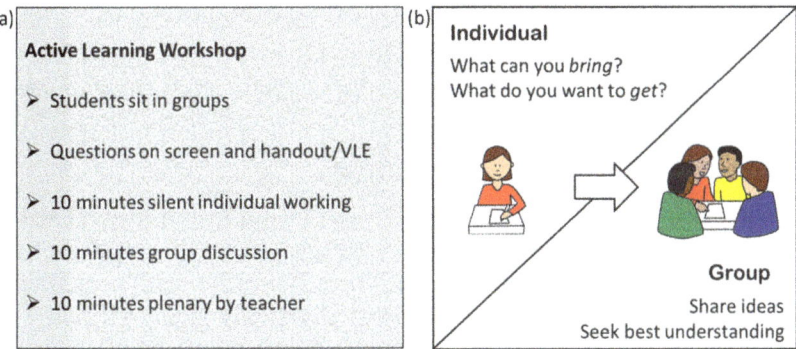

Figure 3.6. (a) Suggested method for active learning workshop. (b) Cartoon to explain method to students. (b) Adapted from https://openclipart.org/detail/227552/student-writing and https://openclipart.org/detail/271246/diverse-student-group. Images stated to be in the public domain.

think and attempt all questions without being interrupted or influenced by others. (See section 3.5 on learning preferences for further explanation of this quiet time.) During this time students can write down anything they think is relevant to the questions: laws, equations, diagrams, facts, information about similar situations or ideas from personal experience, and try to solve all questions. They can access their notes since such an active learning session is primarily about understanding, not memory. It is also helpful for students to make a note of what they don't know, or are unsure of, so that when the group discussion starts, students know what they are looking for. This makes for more fruitful and focussed peer discussion. For students who are very confident in their understanding of the questions, this individual part is not wasted time, but a chance for them to formulate a good 'argument' or convincing explanation to explain their responses to the group during the discussion phase.

It is helpful to explain this procedure in advance to students (see figure 3.6(b)) so they know what to expect, since clarity and structure in instruction have been shown to build positive emotions (Frenzel *et al* 2007). The individual working should allow all students to have something to *bring* to the discussion, and know what they want to *get* from the discussion. However much or little they can offer it is good for students to enter the discussion understanding their learning needs, and having confidence to offer something to the conversation.

To make the most of the peer discussion phase it is good if the group tries to produce a written justification for their choice of answer to each question. This encourages really deep discussion and can help students reach a consensus if required. An example of such a justification is given in figure 3.7, and it is advisable to work through such an example with students in advance so they know the sort of 'evidence' they need to bring to their answer.

This sort of justification, using known knowledge and manipulation to 'prove' the answer for the given situation, really helps students (and teachers) to know if a deep understanding has been reached (Ballard 2017, Carnegie Learning 2018, Williams and Lombrozo 2013, Kuhn *et al* 1988). It is in writing out this evidence that students really see the patterns, implications and links within and between topics of the discipline.

Finally after the individual and group working activities, it is important to have a plenary, where the teacher can draw ideas together with the whole class. This clarifies everyone's thinking, lets students see a succinct justification of the solution, and rounds off the session leaving correct ideas uppermost in students' minds.

3.4 Room layout and group formation

To facilitate peer discussion, it is obviously helpful if students can easily see and hear each other. To this end it is good if rooms with small movable tables can be used, so students within a group can face each other, to listen and contribute to the conversation. Indeed many institutions have developed high tech active learning environments, and Talbert and Mor-Avi (2019) give a detailed review of such spaces. But with large classes such layouts are not always possible, as available rooms can be too small for the whole class. However, fruitful peer discussion can still be held in traditional lectures theatres, and indeed when PI was first proposed this

> **QUESTION**
>
> Which of the following (on its own) would double the force between two electric charges, Q_1 and Q_2 separated by distance r in air?
>
> 1. doubling the charge of Q_1
> 2. halving the distance between the charges
> 3. filling the gap between the charges with plastic.
>
> | A | B | C | D | E |
> |---|---|---|---|---|
> | 1 only | 3 only | 1 and 2 | 2 and 3 | 1, 2 and 3 |
>
> **JUSTIFICATION**
>
> Electrostatic force is given by Coulomb's law: $F = \dfrac{kQ_1Q_2}{r^2}$
>
> 1. $F_1 = \dfrac{k2Q_1Q_2}{r^2} = 2F$ *true*
>
> Multiplying one charge by 2, doubles the force
>
> 2. $F_3 = \dfrac{kQ_1Q_2}{(r/2)^2} = 4F$ *false*
>
> Since the radius is squared, and in the denominator, reducing r by factor of 2, increases the force by factor of 2^2, i.e. by 4
>
> 3. Plastic is a dielectric, and will reduce the force between the charges. *false*
>
> **ANSWER:** just statement 1 is true, therefore answer is A.

Figure 3.7. Showing the type of evidence that students need to bring to their justification.

was how it was implemented (Mazur 1997). Even in serried rows students can discuss with those close to them (maybe either side, and in front or behind). In this situation it is pragmatic to ask students to leave every third row free if possible, so the teacher can walk amongst them. Whatever layout is used, I myself have found it really valuable to walk around and listen to how students are debating the different ideas, and this helps to inform my plenary session with them. Research by Vercellotti (2017) investigated the difference between modern interactive learning spaces and traditional classrooms, concluding that there is no difference in students' learning, but that interactive classrooms can make the learning process more effective and efficient.

It is also worth considering the optimum size of group for effective peer discussion. Whilst there is no correct size and student interactions depend on many factors, research by Akinola and Ayinla (2014) and Corrégé and Michinov (2021) concluded that the most effective group size is four. Whilst larger groups allow more and varied views to be surfaced, the larger the group the less interaction per member. Smaller groups can be very effective but can also result in discussion being a little strained or one-sided if students do not have much to bring to the discussion.

Looking at group formation Smith *et al* (2018) have studied the effect on learning achievement of whereabouts students sit in the lecture theatre. They found that students typically sit with their friends, and that the different friendship groups achieved similar levels of attainment in problem solving tasks. However, students who sat alone tended to perform less well in these tasks. They concluded that engagement and attainment are strongly related to inclusion in a peer group, and these factors are considered further in section 3.5 below.

3.5 Learning preferences

All students are different and learn in different ways, and sometimes we can be guilty of thinking that the research on active learning means that everyone benefits from talking, discussion and group interactions. Whilst it is helpful to be challenged to think more deeply about their discipline by comments (correct or incorrect) from their peers, for some students group work is an extremely uncomfortable and unsettling experience, and as teachers we need to find ways to achieve active learning without pressurizing students to participate in a traumatic experience.

The sorts of students we particularly need to consider here are neurodivergent students. Neurodiversity includes things like autism, ADHD and dyslexia. Many neurodivergent students find it difficult to communicate their thinking and can be overwhelmed by sensory inputs, and thus peer discussion can cause them extreme anxiety. This is particularly important in STEM disciplines which have a higher proportion of neurodivergent students than in the general student population (Wei *et al* 2013).

In fact, Lawrence (2016) describes many STEM students as being 'introverted' in their approach to learning. This is not that they are considered shy, but that they think more slowly, prefer quiet and solitude away from distracting noise and clutter. They like to formulate and practise their responses before discussing with others, and prefer one-on-one conversations to large groups. This is in contrast to students who are 'extrovert' in their approach to learning. Such students become energized by others, and can translate thoughts into speech and make decisions at speed.

So we need to acknowledge that some of our students will thrive in group work, others will benefit even if feeling a little out of their comfort zone, but some will find it too overwhelming to really engage. We thus need to separate the elements of active learning from the typical deployment methods utilizing group discussion. Active learning involves students in thinking and being challenged to understand the workings, patterns, implications and links between aspects of the subject. This can be a solitary activity if needed, if the questions are posed well and scaffolded.

However, it would be better to make the peer discussion activity more welcoming to such students. The 10 min of individual thinking proposed above in the workshop method (section 3.3) has been devised with such students in mind, to have quiet reflective time where all can think and formulate what they want to say, and identify what they want to get out of the group work, with no distractions in the room. Further useful advice can be obtained from the National Autistic Society and work by Honeybourne (2018).

Before the peer discussion starts it helps to have a collective understanding between the teacher and all students that there is no pressure to speak. Students can learn a lot just from listening to others challenge and debate the nuances of the subject. All students should be invited to speak in the group, so they feel welcome, but it should be acceptable if they decline to speak. For some students even sitting in a group is too much, and it should be considered as 'normal' if they decide to sit on the fringes of the room, working alone, but maybe listening in to others' conversations.

3.6 Concluding remarks

In this chapter we have covered the practical elements for developing active learning in large STEM classes. We have seen that posing conceptual questions with peer discussion encourages students to think more deeply about what they are studying, rather than just accepting new information and learning by rote. Furthermore, we have seen that asking students to justify their answers pushes them to pursue understanding rather than just seek to obtain the right answer. A variety of question styles have been presented, showing how they address different cognitive processes, along with ideas for sourcing or creating such questions appropriate to the material being taught.

Practical suggestions have been given for deploying active learning with different room layouts, from moveable tables to fixed serried rows. We have seen the benefit of setting aside a period for individual quiet thinking ahead of peer discussion, to provide an inclusive learning environment which particularly supports neurodiverse students, and allows all students to assimilate what they already know and identify where they need further clarification. The formation of peer groups has been further discussed, to ensure that all students feel comfortable and able to participate in the session. So in summary, whilst the exact methods of active learning can be chosen to suit the situation, the subject and the students, it is the combination of carefully created conceptual questions and the opportunity for students to listen and explain these concepts to others that fosters deep learning.

References

Akçayır G and Akçayır M 2018 The flipped classroom: a review of its advantages and challenges *Comput. Educ.* **126** 334–45

Akinola O S and Ayinla B I 2014 An empirical study of the optimum team size requirement in a collaborative computer programming/learning environment *J. Softw. Eng. Appl.* **7** 1008–18

Anderson L and Krathwohl D 2001 *Taxonomy for Learning, Teaching, and Assessing, A: A Revision of Bloom's Taxonomy of Educational Objectives* ed P W Airasian, K A Cruikshank, R E Mayer, P R Pintrich P, J Raths and M C Wittrock M (New York: Longman)

Ballard D 2017 Discourse in math—don't just talk about it *Consortium on Reaching Excellence in Education* https://corelearn.com/wp-content/uploads/2017/08/discourse-in-math-whitepaper.pdf#:~:text=Mathematical%20discourse%20is%20the%20verbal%20and%20written%20communication,synthesize%20it%20and%20to%20retain%20it%E2%80%9D%20(p.%2021%29 (accessed 3 March 2023)

Berk R A 1996 A consumer's guide to multiple-choice item formats that measure complex cognitive outcome untitled pearsonschool.com (accessed 4 March 2023)

Bloom B S 1956 *Taxonomy of Educational Objectives: The Classification of Educational Goals. Handbook 1: Cognitive Domain* ed M D Engelhart, E J Furts, W H Hill and D R Krathwohl (New York: David McKay)

Carnegie Learning 2018 Why students need to explain their reasoning https://carnegielearning.com/blog/why-students-need-to-explain-their-reasoning/ (accessed 3 March 2023)

Chen Y, Wang Y, Kinshuk and Chen N-S 2014 Is FLIP enough? Or should we use the FLIPPED model instead? *Comput. Educ.* **79** 16–27

Chin C 2007 Teacher questioning in science classrooms: approaches that stimulate productive thinking *J. Res. Sci Teach* **44** 815–43

Corrégé J-B and Michinov N 2021 Group size and peer learning: peer discussions in different group size influence learning in a biology exercise performed on a tablet with stylus *Front. Educ.* **6** 733663

Crouch C H and Mazur E 2001 Peer instruction: ten years of experience and results *Am. J. Phys.* **69** 970–7

Crouch C, Fagen A, Callan J and Mazur E 2004 Classroom demonstrations: learning tools or entertainment? *Am. J. Phys.* **72** 835–8

Cummins S, Beresford A R and Rice A 2016 Investigating engagement with in-video quiz questions in a programming course *IEEE Trans. Learn. Technol.* **9** 57–66

Entwistle N 1991 Approaches to learning and perceptions of the learning environment *High. Educ.* **22** 201–4

Farabaugh M 2020 *Using Student Misconceptions as a Guide for Writing Multiple-Choice Items for AP Chemistry* Chemical Education Xchange chemedx.org (accessed 4 March 2023)

Freeman S, Eddy S L, McDonough M, Smith M K, Okoroafor N, Jordt H and Wenderoth M P 2014 Active learning increases student performance in science, engineering, and mathematics *Proc. Natl. Acad. Sci.* **111** 8410–5

Frenzel A C, Pekrun R and Goetz T 2007 Perceived learning environment and students emotional experiences: a multilevel analysis of mathematics classrooms *Learn. Instr.* **17** 478–93

Honeybourne V 2018 *The Neurodiverse Classroom—A Teacher's Guide to Individual Learning Needs and How to Meet Them* (London: Jessica Kingsley Publishing)

Johnstone A H and Ambusaidi A 2000 Fixed response: what are we testing? *Chem. Educ. Res. Pract. Eur.* **1** 323–8

King A 1993 From sage on the stage to guide on the side *Coll. Teach.* **41** 30–5

Kothiyal A, Majumdar R, Murthy S and Iyer S 2013 Effect of think-pair-share in a large CS1 class: 83% sustained engagement *Proc. of the Ninth Annual Int. ACM Conf. on Int. Computing Education Research*

Kuhn D and Udell W 2003 The development of argument skills *Child Dev.* **74** 1245–60

Kuhn D, Amsel E and O'Loughlin M 1988 *The Development of Scientific Thinking Skills* (San Diego, CA: Academic)

Lasry N, Charles E and Whittaker C 2016 Effective variations of peer instruction: the effects of peer discussions, committing to an answer, and reaching a consensus *Am. J. Phys.* **84** 639–45

Lawrence W K 2016 *Personality and Prejudice* (Washington DC: Paramount Education)

Lin S-Y and Singh C 2013 Can free-response questions be approximated by multiple-choice equivalents? *Am. J. Phys.* **81** 624–9

Lyman F 1981 The responsive classroom discussion ed A S Anderson *Mainstreaming Digest* (College Park, MD: University of Maryland College of Education) pp 109–13

Makhene A 2019 The use of the socratic inquiry to facilitate critical thinking in nursing education *Health SA Gesondheid* **24** a1224

Mazur E 1997 *Peer Instruction: A User's Manual* (Upper Saddle River, NJ: Prentice-Hall)

Nash J M and Krauss K E M A 2015 Method for aligning MCQ assessment with cognitive skills and learning objectives *44th Conf. of the Southern African Computer Lecturers' Association, Johannesburg, South Africa (London)* (National Autistic Society) (accessed 3 March 2023) autism.org.uk

Novak J D 2002 Meaningful learning: the essential factor for conceptual change in limited or inappropriate propositional hierarchies leading to empowerment of learners *Sci. Educ.* **86** 548–71

Piaget J and Inhelder B 1969 *The Psychology of the Child* (London: Routledge and Kegan Paul)

Porter L, Bailey Lee C, Simon B and Zingaro D 2011 Peer instruction: do students really learn from peer discussion in computing? *Conf. Proc. ICER11: Int. Computing Education Research Workshop*

Prince M 2004 Does active learning work? A review of the research *J. Eng. Educ* **93** 223–31

Rodgers K J, Diefes-Dux H A, Cardella M E and Fry A 2012 First-year engineering students' peer feedback on open-ended mathematical modelling problems *Proc.—Frontiers in Education Conf.*

Shaibar H S and van der Vleuten C P M 2013 The validity of multiple choice practical examinations as an alternative to traditional free response examination formats in gross anatomy *Anat. Sci. Educ.* **6** 149–56

Simon and Snowdon S 2014 Multiple-choice vs free-text code-explaining examination questions *Proc. 14th Koli Calling Int. Conf. on Computing Education Research* pp 91–7

Smith D P, Hoare A and Lacey M L 2018 Who goes where? The importance of peer groups on attainment and the student use of the lecture theatre teaching space *FEBS Open Bio.* **8** 1368–78

Smith M, Wood W, Adams W, Wieman C, Knight J, Guild N and Su T 2009 Why peer discussion improves student performance on in-class concept questions *Science* **323** 122–4

Talbert R 2017 *Flipped Learning: A Guide for Higher Education Faculty* (Sterling, VA: Stylus Publishing LLC)

Talbert R and Mor-Avi A 2019 A space for learning: an analysis of research on active learning spaces *Heliyon* **5** 1–19

Thomas P R and Bain J D 1984 Contextual dependence of learning approaches: the effects of assessments *Hum. Learn.* **22** 227–40

Vercellotti M L 2017 Do interactive learning spaces increase student achievement? A comparison of classroom context *Act. Learn. High. Educ.* **19** 197–210

Vickrey T, Rosploch K, Rahmanian R, Pilarz M and Stains M 2015 Research-based implementation of peer instruction: a literature review *CBE—Life Sci. Educ.* **14** es3

Vygotsky L 1978 Interaction between learning and development *Readings in the Development of Children* ed M Gauvain and M Cole (New York: Scientific American Books) pp 34–40

Wei X, Yu J W, Shattuck P, McCracken M and Blackorby J 2013 Science, technology, engineering, and mathematics (STEM) participation among college students with an autism spectrum disorder *J. Autism Dev. Disord.* **43** 1539–46

Williams J J and Lombrozo T 2013 Explanation and prior knowledge interact to guide learning *Cogn. Psychol.* **66** 55–84

Wilson K and Fowler J 2005 Assessing the impact of learning environments on students' approaches to learning: comparing conventional and action learning designs *Assess. Eval. High. Educ.* **30** 87–101

IOP Publishing

Effective Teaching in Large STEM Classes

Anna K Wood

Chapter 4

Using classroom observation tools to characterize large classes

Anna K Wood and George Kinnear

Classroom observation tools give detailed data about the activities that take place in learning environments. They are particularly useful for understanding what happens in large STEM classes. This chapter will discuss the different tools available and their relative merits. It will then provide a detailed guide to using the FILL+ tool (Framework for Interactive Learning in Lectures) and to interpreting FILL+ data. Finally, the chapter will explore some of the findings in the literature based on using classroom observation tools, and discuss the future potential for these tools.

4.1 What are classroom observation tools? Why do we need them?

With the growing understanding that using active and interactive engagement techniques is beneficial for learning in large classes (see chapter 2), has come the realization that we need good quality data about what happens in classroom environments. To address this need, a range of classroom observation tools have been developed, each with different characteristics. These tools can give a detailed picture of what classes look like: how much time is spent on each type of activity, how the different activities are used throughout a class period, and what the students and the lecturer spend time doing during the class.

The ultimate goal of these tools is to improve the quality of teaching, particularly in large classes. There are two main ways in which these tools can help toward this goal. First, they provide education researchers with a way to characterize teaching that can be used to compare approaches across many classrooms (e.g., Stains *et al* 2018) and to understand the effectiveness of different approaches (e.g., Maciejewski 2016). Second, they can be used to provide feedback to teachers, as a way to help them reflect on their teaching and identify professional development needs.

In this chapter we give an overview of the different characterization tools that have been developed, and discuss the advantages and disadvantages of each one.

Our focus here is on how these tools help us to understand large STEM classes (rather than other forms of learning environments such as workshops). We provide a guide to using the FILL+ tool (Framework for Interactive Learning in Lectures) and to interpreting FILL+ data. We focus on FILL+ since it provides a good balance between detail about classroom activities and ease of use, while also giving accurate, second-by-second information about the length of time spent on different activities. Finally we explore what classroom observation tools tell us about teaching in large STEM classes, how they can support teacher development and the potential uses of these tools to improve teaching in large classes.

4.2 Summary of characterization tools

Approaches to characterizing classroom activities fall into two main types: those that require judgements to be made about some aspect of the teaching and those that aim, as far as possible, to be non-judgemental in that they make no deductive assumptions about which teaching practices may be more or less effective. An example of the former is the Reformed Teaching Observation Protocol (RTOP) (Sawada *et al* 2002). RTOP asks observers to make judgements about the teaching, specifically about how well the teaching conforms to a predetermined notion of what learner-centred teaching looks like. RTOP is used to evaluate teaching, and provides a score which relates to how 'reformed' the teaching is deemed to be. Similarly, the teaching practices inventory (TPI) (Wieman and Gilbert 2014) asks teachers to fill in a survey covering different aspects of their teaching, such as the learning outcomes, the materials they provide, how much time they think they spend talking and how many times the students interact with each other. This results in an ETP score (extent of use of research-based teaching practices) which, as the name suggests indicates how closely the teaching practice aligns to what the authors deem to be best practice. TPI requires the teacher to score themselves and relies on their memory of what has taken place in the classes.

In contrast, a number of tools have been developed that aim to produce a non-judgemental characterization of what takes place in the classroom. These generally set out a detailed list of activity types, so that observers can record which types of activity are taking place at different times. These include COPUS and FILL+. This chapter focuses on a selection of the most commonly used characterization tools. These are shown in table 4.1 along with their key features, including the number of codes, the timeframe over which observations are made, whether multiple codes are possible at any given time, and the time needed for training.

4.2.1 TDOP

The teaching dimensions observation protocol (TDOP) (Hora *et al* 2013) was designed as a research instrument to analyse post-secondary teaching. It defines 46 codes over six dimensions, covering all aspects of both teacher and student behaviour. TDOP is coded in 2 min intervals. However, due to the complexity it requires substantial training to achieve good inter-rater reliability.

Table 4.1. Summary of classroom observation tools.

Tool name	No. codes	Timeframe for observations	One code at a time	Time needed for coding per minute	Training needed	Reference
TDOP	46	2 min	No		28 h[a]	Hora et al (2013)
COPUS	25	2 min	No	1 min 20 s[b]	1.5–25 h[b]	Smith et al (2013)
PORTAAL	21	1 s	Yes	1 min 48 s[b] 2 min 30 s[c]	5–20 h[b]	Eddy et al (2015)
FILL/FILL+	10	1 s	Yes	1 min 10 s[c]	6–8 h[c]	Wood et al (2016)/ Kinnear et al (2021)
DART	3	0.5 s	Yes		2.5[b]	Owens et al (2017)

[a] Hora et al (2013)
[b] Asgari et al (2021)
[c] Kinnear, unpublished

4.2.2 COPUS

The Classroom Observation Protocol for Undergraduate STEM (COPUS) (Smith et al 2013) was developed to provide a simpler tool that takes less training than TDOP. COPUS has 25 codes in two categories: what the students are doing and what the instructor is doing. COPUS, as with TDOP, provides data about classroom activities in 2 min time segments. The authors of COPUS claim that observers can be trained in only 1.5 h, but other research teams reported spending up to 8 h (Kinnear, unpublished) and 25 h (Asgari et al 2021) to attain high inter-rater reliability. COPUS is the most commonly used classroom observation tool in the US.

4.2.3 PORTAAL

PORTAAL (Practical Observation Rubric to Assess Active Learning) (Eddy et al 2015) uses continuous coding rather than 2 min intervals like COPUS and TDOP. It has 21 codes which cover four dimensions of best practice in active learning: practice, logic development, accountability and apprehension reduction. However, most behaviours are counted as frequencies rather than duration. The main challenge for novice coders, similar to COPUS, is categorizing the cognitive sophistication of questions asked by lecturers (Chinnery et al 2018).

4.2.4 FILL/FILL+

The Framework for Interactive Learning in Lectures (FILL) (Wood et al 2016) focuses on the types of interactions that take place in large classes, particularly those employing peer instruction. FILL+ (Kinnear et al 2021) is a development of FILL which is more generalizable to large classes that use a variety of different active learning approaches. Both FILL and FILL+ provide a continuous (i.e., second by second) analysis. FILL+ has a total of 10 codes and is designed to be carried out from lecture recordings. The coded activities are mutually exclusive and therefore only one activity is coded at any one time. The codes in FILL+ aim to be self-explanatory so that it is easy for someone new to the tool to use accurately. This contrasts with PORTAAL where some elements, such as coding the type of questions being used, require judgements to be made. FILL/FILL+ are the most commonly used protocols in the UK. Training to achieve a good inter-rater reliability takes approximately 6–8 h.

4.2.5 DART

The DART (Decibel Analysis for Research in Teaching) tool (Owens *et al* 2017), is a system for automatically analysing classroom activities. It works by determining sound levels from the classroom which are then translated into one of three types of activity: no-one talking (students working individually), one person talking (normally lecturer talking) and multiple voices (student discussion). This means that the level of detail is much more limited than the protocols described above. However, it does not rely on humans for the analysis, and is much cheaper and quicker to use.

4.2.6 Other tools

The genre of classroom observation tools has been expanded beyond the lecture environment. For example, Velasco *et al* (2016) developed a protocol for characterizing activities in chemistry labs—the Laboratory Observation Protocol for Undergraduate STEM (LOPUS) and Kranzfelder *et al* (2019) have created the Classroom Discourse Observation Protocol (CDOP), a method for characterizing the speech patterns (teacher discourse moves) in undergraduate STEM learning environments.

4.3 Comparison of tools

The choice of tool for characterizing classroom activities will depend on the type of data that you are looking for, the time available to spend on analysing the classes and the level of detail needed. The easiest and least time consuming is the automatic tool DART, however this will only give information about three different types of activities—the lecturer talking, students spending time discussing and students spending time engaged in problem solving alone. This will give a good estimate of how much time is spent on active learning (the latter two options), but will not give information on the types of active learning that have taken place. The most notable activity that is missing is anything related to whole-class discussions, such as questions to and from the lecturer. However, Asgari *et al* (2021) found that the information from DART was the easiest to understand for teachers with the least time investment compared to COPUS and PORTAAL.

Both COPUS and PORTAAL have a large number of observation codes (25 and 21 respectively), which means they give a very detailed picture of what is happening. In comparison FILL+ has a modest 10 codes. However, the greater complexity of COPUS and PORTAAL means that they take longer to learn, and they produce data that is more complex to understand. With COPUS, it is viable to use two codes in tandem, whereas FILL+ specifies that only one code can be used at a time.

Another key difference is the duration of the time samples: for FILL+ and PORTAAL coding is conducted on a second-by-second basis, whereas in COPUS and TDOP activities are coded using 2 min segments. This means that data from FILL+ is more accurate, as it gives the time spent on activities to the nearest second. This is particularly relevant for activities with a short time span, such as asking and answering questions. We have recently analysed a set of classes using both COPUS and FILL+, enabling a comparison of the results from the two tools. While COPUS showed 30% of the 2 min blocks contained questions to and from the lecturer, FILL+ revealed that, in fact, only 3% of the time was spent on questioning for the same set of lectures. Similarly, 91% of the 2 min blocks contained COPUS codes representing the lecturer talking, whereas the FILL+ analysis showed the actual time spent on lecturer talk was 75% of the class. Thus, COPUS appears to overestimate the time spent on activities compared to FILL+.

4.4 Guide to using FILL+ (and interpreting the data)

This section provides a detailed guide to using FILL+ and interpreting FILL+ data. FILL+ has been chosen as it provides a good balance between complexity and ease of use, while also providing accurate data on the time spent on different classroom activities. FILL+ focuses on the way in which students experience the class, with particular focus on the interactions that they take part in—whether those interactions are with other students, with the lecturer, or with the material. As discussed in chapter 2, it is interactions (with others and with the material) that are the essential components of active learning (Murphy and Sharma 2010).

FILL+ builds on the original FILL protocol (Wood *et al* 2016) which was developed to analyse teaching in large physics classes usings a flipped approach combined with peer instruction (Mazur 1999). In this initial implementation of the FILL protocol, two sets of codes were developed: one which referred to the type of activity that took place during the class and one which referred specifically to the different stages of peer instruction. In order to make the FILL protocol more generalizable to classes which use different types of active learning approaches, Kinnear *et al* (2021) developed FILL+ which consists of 10 codes (see figure 4.1). These codes cover all the main activities that take place in large classes, such as the lecturer talking (LT), the students thinking about a problem by themselves (ST), students discussing with each other (SD) and questions from the students (SQ) and the lecturer (LQ) as well as the responses to questions (LR and SR, lecture response and student response, respectively). For a detailed explanation of each of the codes see the FILL+ manual (Smith *et al* 2020).

As shown in figure 4.1, each code of FILL+ is also assigned one of three interactivity levels: **interactive**, **non-interactive** and **vicarious interactive.** The level of interactivity is important, because it is interactions that are the key components of active learning. The interactive category covers codes such as students discussing in small groups (SD) or students thinking about a problem individually (ST) which

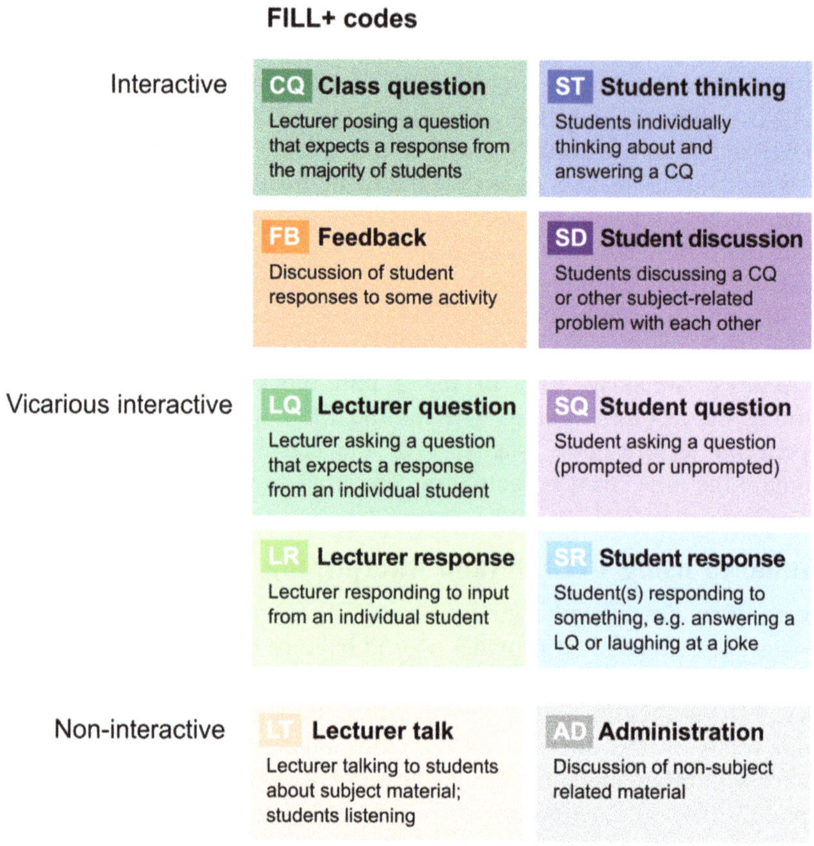

FILL+ codes

Interactive

CQ Class question
Lecturer posing a question that expects a response from the majority of students

ST Student thinking
Students individually thinking about and answering a CQ

FB Feedback
Discussion of student responses to some activity

SD Student discussion
Students discussing a CQ or other subject-related problem with each other

Vicarious interactive

LQ Lecturer question
Lecturer asking a question that expects a response from an individual student

SQ Student question
Student asking a question (prompted or unprompted)

LR Lecturer response
Lecturer responding to input from an individual student

SR Student response
Student(s) responding to something, e.g. answering a LQ or laughing at a joke

Non-interactive

LT Lecturer talk
Lecturer talking to students about subject material; students listening

AD Administration
Discussion of non-subject related material

Figure 4.1. Summary of the codes used in FILL+ to describe the activities taking place in a lecture.

involve substantial levels of interactions either with each other or with the material. The non-interactive category primarily refers to the lecturer talking (and therefore the students listening). Finally, the vicarious interactive category describes activities where there is some interaction between a student and the lecturer, such as questions to and from the lecturer. We felt that such interactions do not fall into the fully interactive category as most students are not themselves engaged in the interaction; however, students not directly engaged in the dialogue are still likely to 'follow along', perhaps thinking of their own questions or answers. In this way, such episodes feel substantially different to listening to a monologue, so do not count as non-interactive.

There are a number of key features which set FILL+ apart from other classroom observation tools:

(1) FILL+ has only one possible behaviour code at any given time with the main focus on how students are experiencing the class. While some other coding systems such as COPUS will give a separate code for what teachers are doing and what students are doing, in FILL+ this is built into a single

code. For example, a code of LT (lecturer talking) assumes that students are listening/making notes. Similarly, a code of SD (student discussion) assumes that the lecturer is not interacting with significant numbers of students.
(2) FILL+ is coded on a continuous (second by second) basis which gives an accurate representation of the time spent on different activities.
(3) The 10 possible codes give enough variation to describe the different types of activity while keeping the coding simple enough that coders can remember the codes without needing to refer to a manual. This means that it is straightforward to train coders (for example, undergraduate project students) to code accurately with FILL+.

4.5 Coding with FILL+

FILL+ was designed to be used with lecture captures (lecture recordings) so that the observer does not need to be present in the live class. This also means that multiple observers can code the same lecture to check for agreement (inter-rater reliability). We have also found that good reliability can be obtained using audio-only versions of lecture captures.

For anyone wishing to learn how to code using FILL+, a training manual is available which includes video clips (https://osf.io/2exvh/). We also provide R code for analysing the data (https://osf.io/qm36d). The training manual includes a detailed description of the 10 FILL+ codes shown in figure 4.1, and a link to a short clip of a lecture with and without commentary, to help with familiarization with the codes. There are also links to three more examples with a template coding response, and notes on the reasons why the codes were selected. Once the coder has practised with these, they should be proficient in coding with FILL+. Based on previous experience, we estimate that this whole process should take about 6–8 h.

In order to code a lecture, the observer chooses the code for the behaviour which most appropriately reflects the situation and records the time this state is initiated. A new code is then introduced (with a time stamp) when it is clear from the student perspective that a new activity is taking place. The R code is then able to calculate the time spent on each of the activities and produce graphs such as the timelines. We strongly recommend measuring the inter-rater reliability, which gives a measure of how similar the coding decisions are for two (or more) observers. It is a good idea to do this at the beginning of a project to check that all the coders have a shared understanding of the meaning of the codes, and to repeat this with short sections of lectures throughout the project to make sure that coding remains reliable.

There are various approaches that can be taken to compute inter-rater reliability. In the original FILL work (Wood 2016), Cohen's Kappa was used and calculated to be 0.74, which indicates a good measure of reliability. COPUS (Smith *et al* 2013) also used Cohen's Kappa and calculated scores between 0.79 and 0.87. In the FILL+ protocol, Krippendorff's alpha was calculated to be 0.852 (where anything above 0.8 is considered to be a good indicator of reliability). We also used a simpler measure which compared how many seconds the two coders agreed on the code chosen and found agreement of 95.7%.

> **Tips for using FILL+**
>
> - Use a random selection of classes from the course—but avoid the first and last classes which are often different to the majority.
> - Stains *et al* 2018 recommended that at least four classes should be coded from a course to get a representative sample.
> - Compare codes from a few samples of different classes with another coder and calculate an inter-rater reliability score to make sure there is good agreement between coders.
> - Consider offering teachers a chance to discuss the data rather than expecting them to interpret it alone (see below for a discussion of helping teachers to reflect with FILL+ data).

4.6 Understanding FILL+ data

The most detailed form of output from FILL+ is the timeline graphs. Timelines show a temporal distribution of the activities throughout the lecture with different colour blocks representing different codes. Figure 4.2 shows four examples of timelines from teachers in large classes at the University of Edinburgh which illustrates the variation seen in different teaching styles. The top graph is from a class that is predominately teacher-led where the lecturer spent substantial periods of time talking (pale orange) with a few brief interruptions for questions (light and dark green), mostly from the lecturer to the students. The second graph shows a class which has occasional use of interactive engagement techniques but also large periods of time when the lecturer is talking. The bottom two graphs are from classes that have a substantial active learning component, and which are discussed in detail in other chapters of this book. Pamela Docherty (chapter 9) makes use of peer instruction in her classes, and the peer instruction cycle can be clearly observed in the timelines (beginning with a class question, followed by student thinking, student discussion and finally feedback). Heather McQueen uses an approach called 'quectures' (see chapter 10), which includes many whole-class questions and substantial periods of student discussion. In both cases, there is also substantial time spent on questions to and from the lecturer.

Using the FILL+ data, it is also possible to provide simplified timelines based on the level of interactivity, as shown in figure 4.3. This example shows lectures by Ross Galloway, in a follow-on course to the one discussed in chapter 7. The detailed timelines show instances of the classic peer instruction cycle, alongside several periods of sustained questions-and-answers. The simplified form of the data was found to be particularly helpful for those who use active learning strategies and whose standard graph is quite complicated. In recent work (Wood *et al* 2022) teachers commented when viewing the data for the first time that the simplified version was helpful as it was easier to remember what the three codes represented, but that the detailed version was useful for in-depth reflection at a later date.

4.7 Examples of how classroom characterization tools are used in practice

This section aims to provide a snapshot of the main uses for classroom observation tools discussed in the literature, rather than a systematic review of the research.

4.7.1 Collecting data on faculty practices

Classroom characterization tools provide useful information about what teaching looks like in different subject areas and in different universities. For example, both FILL+ (Kinnear *et al* 2021) and COPUS (Stains *et al* 2018) have been used for large multidisciplinary studies. FILL+ was used to explore 208 classes taught by 37 lecturers across STEM subjects (mathematics, physics or veterinary medicine) at the University of Edinburgh, UK and COPUS was used to analyse STEM courses in a multi-institutional study across 25 American Universities of over 2000 classes taught by 500 faculty members. The FILL+ study found that, on average, 79.3% ± 19% of lecture time is attributable to lecturer talk; this closely matches the figure of 74.9% ±

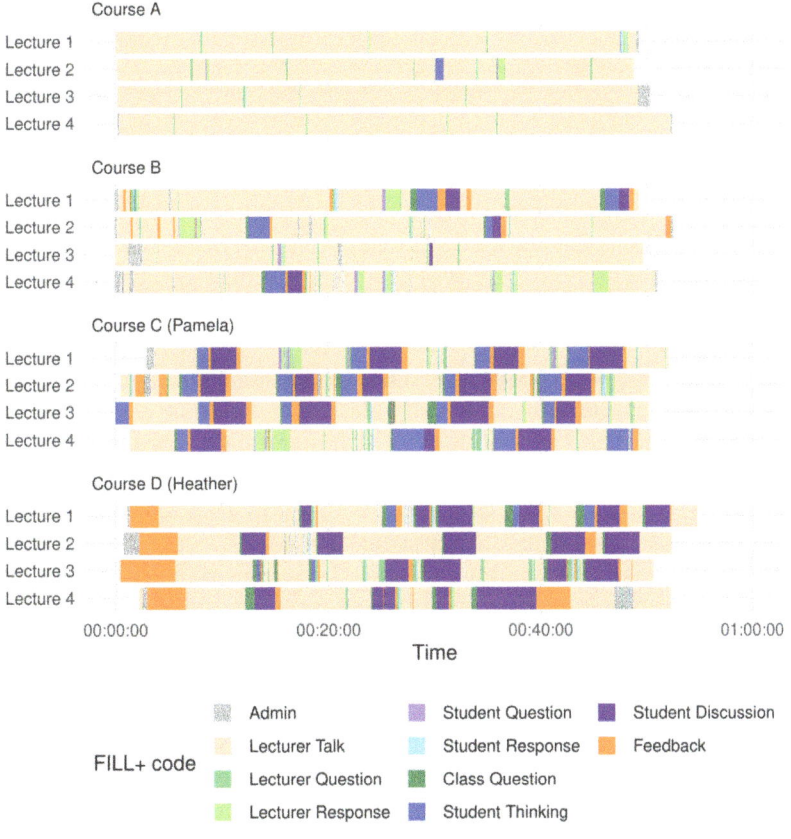

Figure 4.2. Timelines from a FILL+ analysis.

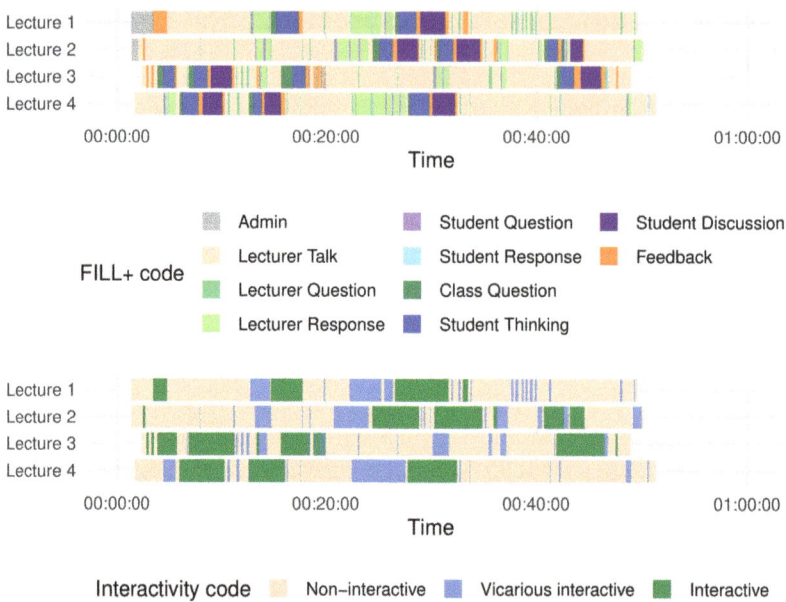

Figure 4.3. Comparison of a full FILL+ timeline (top) with a simplified version showing only the three levels of interactivity (bottom).

27.8% of the total 2 min intervals found using COPUS which were coded as 'lecture'. The FILL+ analysis found some interesting disciplinary differences: mathematics had the smallest time spent on the lecturer talking (71.9±21.2) whereas veterinary medicine had the highest 91.2 (± 9.5). This may reflect the fact that veterinary medicine is a professional course with a large volume of core content, which can result in teachers taking a more didactic approach (Wood *et al* 2022). The COPUS analysis found that biology classes were more frequently associated with the student-centred instructional style, while chemistry classes tended to be classified as didactic.

One important finding from both the FILL+ and COPUS data is that even in classes which are oriented towards active learning, a substantial amount of class time is still taken up by the lecturer talking. Even in the most interactive classes that we analysed using FILL+, the lecturer spent around 50% of the class time talking. Rather than being seen as something negative, it is important to recognize that the lecturer talking can have a range of benefits, especially when used in combination with active engagement strategies. For example, the teacher can model expert thinking while demonstrating how to solve a problem, they can address misconceptions which are evident from the use of clicker questions, and they can respond to gaps in students' understanding with targeted explanations.

The total length of time that the lecturer talks for gives an indication of the teaching approach being used, however it doesn't necessarily give information about how the class is experienced by students. As the FILL+ protocol records the length of time in seconds for each activity, we can also determine the average length of the

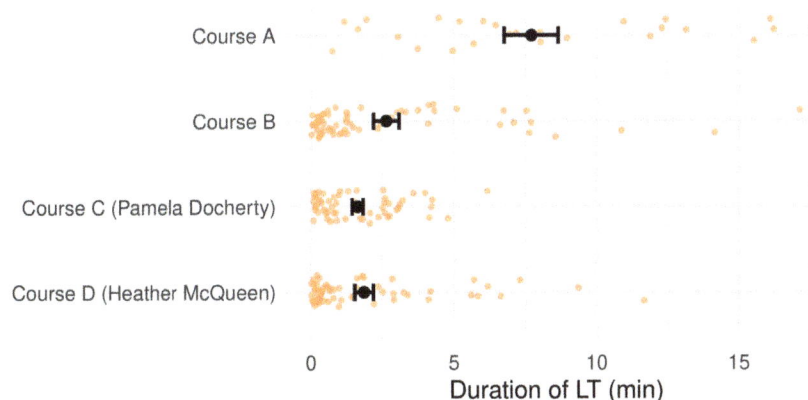

Figure 4.4. Distribution of the duration of lecturer talk (LT) periods, for the examples shown in figure 4.1. Each of the pale orange points shows the duration of a single period of LT. The bold black points and error bars show the mean and standard error.

periods where the lecturer is talking. This can indicate whether the talking is broken up by activities or is a long continuous monologue. Figure 4.4 shows a summary of the duration of every period of lecturer talk, from the same lectures shown in figure 4.2. This shows that in the classes using active learning approaches, it is rare for there to be a period of uninterrupted lecturer talk that is longer than 5 min

4.7.2 Classifying the types of teaching approach used

Data generated from classroom observation tools can be used to characterize the type of teaching approach being used. Stains *et al* (2018) have demonstrated a cluster analysis approach for COPUS data, which found that classes could be assigned to one of three distinct instructional profiles: didactic, interactive lecture, and student centred. A class was determined to be didactic if 80% or more of the 2 min time intervals in the class consist of lecturing. The interactive lecture style was assigned if more than 50% of time intervals contained the lecture code, and if they also incorporated some student-centred activities and group work. The student-centred instructional style is defined by large amounts of student-centred activities and group work taking place in each class period (Stains *et al*, 2018). The authors found that 55% of their data set fitted the didactic instructional profile, 27%, the interactive approach and 18% the student-centred profile.

4.7.3 Gaining greater understanding of pedagogical approaches

Classroom observation tools can be used to give insights into different teaching approaches. For example, work by Maciejewski (2016) compared flipped and non-flipped instances of a calculus course using COPUS. They found that students in the non-flipped sections of the course spent around 83% of the 2 min segments during the class listening to the instructor, compared to 42% of the 2 min segments for the students in the flipped classes. At the end of the course those in the flipped class achieved higher grades on the final exam compared to those in the non-flipped class. They also found that the students with low calculus knowledge (but good mathematical skills) gained the most and that while all students' expert-like attitudes toward the discipline declined over the course, the drop was lower for those in the flipped class. The authors suggest that 'future studies should move on from evaluating the effectiveness of classroom flipping in different contexts and uncover the features and mechanisms of flipping that make it effective' (Maciejewski 2016). Classroom observation tools provide a way to gain a more nuanced understanding of which activities (and combinations of activities) are most effective.

Work by one of the authors (AKW) and colleagues (Wood *et al* 2018) also shows the potential of classroom observation tools for understanding specific aspects of teaching practice. We used a FILL analysis to gain insights into teacher–student dialogues and student questioning during two flipped physics classes that use peer instruction. We found that the majority of the interactions were in the triadic IRF (initiation, response, feedback) form and that dialogues supported learning through involving students in sense-making, guided expert modelling and wonderment questions. Lecturers' use of questions was also investigated by Kinnear *et al* (2021), using FILL+ as a starting point to identify 1620 questions asked by mathematics lecturers across a sample of 136 lectures. We applied an existing coding scheme (Paoletti *et al* 2018) to classify the types of questions asked, and confirmed previous findings that mathematics lecturers tended to ask factual questions, or prompt students to recommend the next step in a proof or example.

Galloway (2018) analysed 49 lectures from two physics courses, one of which was taught with a flipped approach and one which used a more traditional, teacher-centred approach. Each course was taught by two different lecturers. As expected, the course with the flipped approach had substantially more time on interactions compared to the non-flipped course. However, the most interesting finding was between the two lecturers teaching the flipped course. Both made substantial use of peer instruction which can be seen in the FILL analysis, but one also included a substantial amount of informal dialogues though questions to and from the students (vicarious interactions). This highlights the power of classroom observation tools to uncover nuanced detail about what happens in the classroom.

4.7.4 Supporting individual teachers to make changes

Academic development programmes often provide training and support to help teachers to adopt more interactive teaching approaches. For change to happen, teachers need to be aware of the way that they are teaching and to take time to reflect on the decisions that they make and the way that they impact on students' learning. Classroom observation tools can help this process by providing data about what happens in the classroom. Reisner *et al* (2020) provide a guide for chemistry teachers who have been given COPUS data; the guide aims to help teachers to understand

and engage with the results in a way that supports them to make meaningful changes to their teaching.

However, the provision of such data does not necessarily mean that teachers will know how to use this data to reflect productively. This is particularly the case for STEM teachers, who often find the concept of reflection confusing and vague (Cornford 2002). To support teachers to reflect on their teaching, we developed an approach which combined the provision of accurate data about classroom practices (from a FILL+ analysis) with the opportunity to have a meaningful discussion about their data with someone who was not connected to their teaching area (Wood *et al* 2022). Such conversations (termed 'educational development interviews') encouraged the teachers to talk about their reaction to the data, to think about what ways (if any) it might help them to reflect on their teaching, and whether the data might trigger them to change anything about their teaching.

Using this approach, we found that most teachers found the data useful and interesting and that many felt it would help them to reflect on their teaching. Reflection was triggered when the teachers found the data surprising or unexpected, for example when they noticed that they spent a large chunk of time talking without any interactions. Some of the teachers also indicated that they would make changes as a result of seeing their data. Two types of changes were mentioned: using more active learning and spacing the active learning that they already used more effectively to avoid long periods of just talking. These conversations helped the teachers to think about their beliefs about teaching, the reasons for their choices, and ultimately encouraged deep engagement with the data. Further work is needed to explore if this approach leads to future changes to teaching approaches.

4.8 Future potential

Although classroom observation tools have been around for two decades, their full potential for understanding and supporting change in teaching practices is as yet untapped. There are a range of ways in which classroom observation tools may be used in the future, from peer observation, to supporting teacher reflection and in-depth longitudinal studies.

4.8.1 Peer observation

Peer observation, where someone observes a colleague's teaching and then provides feedback on what they observe, is an important element of ongoing teacher development. Classroom characterization tools have the potential to be used as peer observation tools, but to date very little work has explicitly explored how they can be used to support the peer observation process. In a pilot study six observers used COPUS to provide evaluation on teaching for six faculty members based on a COPUS analysis (Daher *et al* 2018). The participants felt that using COPUS helped facilitate discussion of potentially effective pedagogical strategies. Classroom observation tools could provide a starting point for discussions between teachers about their teaching practices in a way which is non-judgemental.

4.8.2 Supporting teacher change

Classroom characterization tools are a useful source of information for teachers wanting to make changes to their teaching, as they provide detailed data about their teaching approaches. This would be particularly useful for teachers who decide to make substantial changes to their teaching, for example by introducing a flipped classroom approach, but would be just as helpful for monitoring smaller changes such as increasing the interactivity in their classes (e.g., by encouraging questions or setting aside time for student discussions). In such cases a snapshot of teaching taken before and after a change is made could help to provide evidence that there has been an impact on the learning experience.

4.8.3 Supporting professional development programmes

The data from classroom characterization tools could provide evidence of the efficacy of academic development programmes, particularly where the aim is to increase the use of active learning techniques. An analysis could be provided before and after attending such a course in order to explore in what ways (if at all) the programme has encouraged changes to teaching approaches.

As discussed earlier, FILL+ shows potential as a tool, when combined with a discussion about the data, for supporting teacher reflection. Work is underway to explore how this approach can be integrated into established academic development programmes at the University of Edinburgh such as the Edinburgh Teaching Award (EdTA), that leads to professional recognition through the Higher Education Academy, and the Postgraduate Certificate in Academic Practice (PgCAP) for new lecturing staff.

4.8.4 Tracking longitudinal changes to teaching practices

Classroom observation data could also be helpful for tracking how teaching is changing over time, whether that is for an individual teacher, across disciplines or institutions, or even nationally. Such data would be helpful for planning for the needs of future generations of teachers.

4.9 Conclusion

To date much of the research into active learning has been focussed on whether active learning works or not. Classroom observation tools can help to unravel the complexities of how active learning leads to improved learning outcomes. But they also have the potential to help us to move beyond thinking about individual activities, to understanding how different activities can be used together most successfully. For example, we know that students discussing a question in small groups can be beneficial, but to what extent is that activity enhanced by some teacher explanation before or after the activity? Is it more beneficial to have some individual thinking time? One way to address such questions would be with detailed studies of particular classroom episodes. Classroom observation tools provide a means to scale up such studies to consider a wide variety of classrooms. Thus, these tools can provide data which can help us to dig deeper into how active learning strategies can be structured optimally.

References

Asgari M, Miles A M, Lisboa M S and Sarvary M A 2021 COPUS, PORTAAL, or DART *Classr. Obs. Tool Comp. Instr. User's Perspect. Front Educ* **6** 740344

Chinnery S, Hughes K and MacKay J R 2018 The active lecture? Exploring engagement in the veterinary lecture through the PORTAAL tool *VetEd: VetEd Symp. 2018*

Cornford I R 2002 Reflective teaching: empirical research findings and some implications for teacher education *J. Vocat. Educ. Train.* **54** 219–36

Daher T, Babchuk W A, Pérez L C and Arthurs L A 2018 Exploring engineering faculty experiences with COPUS: strategies for improving student learning *2018 ASEE Annual Conf. and Exposition*

Eddy S L, Converse M and Wenderoth M P 2015 PORTAAL: a classroom observation tool assessing evidence-based teaching practices for active learning in large science, technology, engineering, and mathematics classes *CBE—Life Sci. Educ.* **14** ar23

Galloway R 2018 What really happens in active engagement lectures? *PERC* http://per-central.org/perc/2018/detail.cfm?ID=7186

Hora M T, Oleson A and Ferrare J J 2013 Teaching dimensions observation protocol (TDOP) user's manual *Madison Wis. Cent. Educ. Res.* http://tdop.wceruw.org/Document/TDOP-Users-Guide.pdf

Kinnear G, Smith S, Anderson R, Gant T, MacKay J R, Docherty P, Rhind S and Galloway R 2021 Developing the FILL+ tool to reliably classify classroom practices using lecture recordings *J. STEM Educ. Res.* **4** 194–216

Kranzfelder P, Bankers-Fulbright J L, García-Ojeda M E, Melloy M, Mohammed S and Warfa A-R M 2019 The classroom discourse observation protocol (CDOP): a quantitative method for characterizing teacher discourse moves in undergraduate STEM learning environments *PLoS One* **14** e0219019

Maciejewski W 2016 Flipping the calculus classroom: an evaluative study *Teach. Math. Its Appl. Int. J. IMA* **35** 187–201

Mazur E 1999 *Peer Instruction: A User's Manual* (Upper Saddle River, NJ: AAPT: Prentice-Hall)

Murphy R and Sharma N 2010 What don't we know about interactive lectures *Int. J. Media, Technol. Lifelong Learn.* **6** 111–9 http://seminar.net/volume-6-issue-1-2010/135-what-dont-we-know-about-interactive-lectures

Owens M T, Seidel S B, Wong M, Bejines T E, Lietz S, Perez J R, Sit S, Subedar Z-S, Acker G N and Akana S F 2017 Classroom sound can be used to classify teaching practices in college science courses *Proc. Natl. Acad. Sci.* **114** 3085–90

Paoletti T, Krupnik V, Papadopoulos D, Olsen J, Fukawa-Connelly T and Weber K 2018 Teacher questioning and invitations to participate in advanced mathematics lectures *Educ. Stud. Math.* **98** 1–17

Reisner B A, Pate C L, Kinkaid M M, Paunovic D M, Pratt J M, Stewart J L, Raker J R, Bentley A K, Lin S and Smith S R 2020 I've been given COPUS (classroom observation protocol for undergraduate STEM) data on my chemistry class… now what? *J. Chem. Educ.* **97** 1181–9

Sawada D, Piburn M D, Judson E, Turley J, Falconer K, Benford R and Bloom I 2002 Measuring reform practices in science and mathematics classrooms: the reformed teaching observation protocol *Sch. Sci. Math.* **102** 245–53

Smith M K, Jones F H, Gilbert S L and Wieman C E 2013 The classroom observation protocol for undergraduate STEM (COPUS): a new instrument to characterize university STEM classroom practices *CBE—Life Sci. Educ* **12** 618–27

Smith S, Anderson R, Grant T and George K 2020 FILL+ training manual online https://doi.org/10.17605/OSF.IO/27863 (accessed 23 June 2023)

Stains M, Harshman J, Barker M K, Chasteen S V, Cole R, DeChenne-Peters S E, Eagan M K, Esson J M, Knight J K and Laski F A 2018 Anatomy of STEM teaching in North American universities *Science* **359** 1468–70

Velasco J B, Knedeisen A, Xue D, Vickrey T L, Abebe M and Stains M 2016 Characterizing instructional practices in the laboratory: the laboratory observation protocol for undergraduate STEM *J. Chem. Educ.* **93** 1191–203

Wieman C and Gilbert S 2014 The teaching practices inventory: a new tool for characterizing college and university teaching in mathematics and science *CBE—Life Sci. Educ* **13** 552–69

Wood A K, Christie H, MacKay J R and Kinnear G 2022 Using data about classroom practices to stimulate significant conversations and aid reflection *Int. J. Acad. Dev.* 1–14

Wood A K, Galloway R K, Donnelly R and Hardy J 2016 Characterizing interactive engagement activities in a flipped introductory physics class *Phys. Rev. Phys. Educ. Res.* **12** 010140

Wood A K, Galloway R K, Sinclair C and Hardy J 2018 Teacher–student discourse in active learning lectures: case studies from undergraduate physics *Teach. High. Educ.* **23** 818–34

IOP Publishing

Effective Teaching in Large STEM Classes

Anna K Wood

Chapter 5

Authentic and inclusive (summative) assessments

Firas Moosvi and Simon Bates

This chapter discusses the design and implementation of authentic and inclusive assessments. It is our (perhaps provocative) position that if the instructional goal is to promote learning, *all* assessments must be formative, and summative assessments should only exist as place-holders until the logistical (and cultural) barriers preventing full adoption of formative assessments can be overcome. We start with an introduction that covers the purpose, and aspirational goal of assessments, followed by general strategies, philosophies and pedagogies for the adoption of authentic assessments, ending with specific implementations to transform traditional summative assessments into more authentic, inclusive, formative assessments for learning.

5.1 Introduction

An authentic assessment is one that establishes connections between real-world experiences and school-based ideas (Lund 1997) through challenging tasks requiring higher-order thinking, problem-solving, and creativity (Archbald 1988). Making assessments inclusive requires a more holistic approach, including consideration of the learning environment, adopting a growth mindset, and implementing more equitable teaching practices (Sathy and Hogan 2022). Practical ways to design inclusive assessments are: removing cultural references and idioms to increase task comprehension, improving access to activities for disabled students, showcasing contemporary and historical figures from equity-deserving groups, and incorporating active learning techniques in the classroom. Before diving into a discussion of designing authentic and inclusive summative assessments, it is worth defining formative and summative assessments, and how they are traditionally disambiguated[1]. Black and Wiliam (1998) proposed a sequence of two actions that

---

[1] We say 'traditionally' because the differentiation is somewhat contentious, and in this chapter we will advocate for all assessments to be formative.

comprise the core of a formative assessment activity: first is the recognition by the learner of a knowledge/skill/understanding gap between their current state and the desired goal and second is the action taken by the learner to close that gap. As strong advocates for formative assessments, Black and Wiliam (2004) note that in general, 'a formative assessment has as its prime purpose the promotion of learning, that is, the first priority in its design and practice is to serve the purpose of promoting students' learning'. This is contrasted with summative assessments, whose primary purpose is to determine 'accountability, ranking, or competence' amongst students, usually for administrative or logistical purposes (Black and Wiliam 2004). Within the context of a course, summative assessments are given at logical endpoints of units/blocks/modules as midterm tests or quizzes, or at the end of a course in the form of final exams. Formative assessments come in many forms and are typically done in low-stakes, un-timed modes with feedback given to students so they can grow and improve. Importantly, even frequent summative testing has formative properties and is a form of distributed practice with learning benefits compared to infrequent summative assessments (Fitch *et al* 1951).

Examining the origins of historical examinations that have morphed into today's closed book pen and paper final exams is well outside the scope of this work (we recommend the review by Broadfoot and Black (2004) as an excellent starting point). Suffice it to say, assessments have a long and sordid history, but it is undisputed that assessments looked very different in the 16th century[2], where university examinations were conducted in public, orally, in Latin, and with spectators and participants from the broader academic community (Stray 2001). By the late 18th century, a 'different notion of fairness' was developed where the focus was on the individual rather than the group, and it was widely believed that the only fair procedure to compare students was to give them all the same questions. Indeed, with the advent of standardized written exams, the capacity and potential for students to learn different content at different rates, or accounting for alternate starting points, or alternate ways of knowing is almost completely ignored. It has also been noted that this approach to education and learning is imperialistic and colonial (Viruru 2006), similar to how IQ tests (Duckworth *et al* 2011), the ACT, and the SAT are also very problematic (Rosales and Walker 2018).

The summative final exam is typically a comprehensive assessment whose purported aim is to determine the depth and breadth of what a student has learned over the semester. They are normally held during a university-wide 'exam period', when classes are suspended and students are given an arbitrary amount of time to review content. Students then strategize on organizing their time to cram as much information as possible so it can be regurgitated on command (whenever their exams happen to be scheduled). Once the exam is done, students tend to wipe the entire experience from their memory and force themselves to forget the mildly traumatic experience. Unfortunately this *forced forgetting* can also happen *during* an exam, depending on the timing of the cram sessions and how much rest students get before

---

[2] Certainly it was a different time but the authors here hold no fondness for the distant past where educational institutions were elitist, racist, sexist, misogynistic and classist.

their exams (Stalnaker and Stalnaker 1934, Jenkins 2013, Huang *et al* 2016). It is troubling that when the purpose of teachers is to help students learn, by far the most widely used tool to measure that learning—closed book, pen and paper final exams—is so deeply flawed (Williams and Wong 2009). It is incumbent on both teachers and the entire education system to critically examine the purpose and role of traditional final exams, and assessments in general.

5.1.1 Purpose of assessment

Regardless of how one feels about assessments and their purported (versus actual) effects, it is clear that assessments are important and have several acknowledged purposes including 'feedback, reporting, certification, selection, accountability, and national comparison' (Broadfoot and Black 2004). Briefly, we'll describe the norms of the pedagogical, logistical and cultural purpose of assessments in post-secondary institutions. Pedagogically, assessments are used to provide feedback to students, give them opportunities to learn, challenge themselves and extend their zones of proximal development (Vygotsky 1978). Culturally, assessments are used to allay fears of the declining 'quality' of education, dispel the 'moral panic' around institutions, teachers and students achieving arbitrary 'standards' (Leathwood 2005). Notably, the public and politicians are most susceptible to obsessing over students achieving these standards. However, accreditation and certification bodies also rely on assessments to determine which degree programs are acceptable, and institutions use assessments to determine which courses comprise these accredited programs. Logistically, assessments have become an essential component of multi-year degree programs and course sequences because they help teachers assign students superficially 'objective' numbers and letter grades. Scholarship, bursaries and grants use final course grades to select high performing students for considerable investment and support. Teachers in senior level courses also rely on very specific performance requirements in lower-level pre-requisite courses, often to control class sizes but usually under the guise of having 'high standards'. This is all notwithstanding decades of work that has demonstrably shown that content knowledge and skill after just a few days away from the content leads to staggering declines in retention and recall (Roediger and Karpicke 2006), especially in courses with high-stakes summative assessments. As higher education institutions grow larger, the love affair with traditional summative assessments has grown to a fever pitch, to the point where some institutions have even (archaically) mandated final exams for all courses at the first- and second-year levels. The pressure to keep summative assessments as the focal point of courses often comes from misplaced notions of fairness and blind faith in blunt tools to surface exceptionalism. For instance, with highly competitive programs and scholarships needing 'objective' ways to sort, filter, rate, and rank students.

5.1.2 Aspirational goal for assessments

For decades educational researchers have been telling us that 'too much emphasis has been placed on the grading function of evaluation, and too little on its role in assisting students to learn' (Crooks 1988). Considering the purposes of formative assessments—learners recognizing gaps and taking action—why would teachers even *want* any assessment they create to be anything but formative? Typical answers usually include some form of pressure from colleagues, upper administration, society and most harmfully, inertia and antiquated conventions to rank, rate and compare students to each other (Kohn and Blum 2020). Naivete and idealism aside, though it is clear these are not good pedagogical reasons for continuing with traditional summative assessments, there are real and tangible factors and barriers preventing the adoption of practices that maintain pedagogical integrity in classes. It is our position that if the instructional goal is to promote learning, *all* assessments must be formative, and summative assessments should only exist as place-holders until the logistical (and cultural) barriers preventing full adoption of formative assessments can be overcome.

So what should be the aspirational goal for a modern assessment? Grant Wiggins wrote that:

> In a truly modern assessment, the challenge is to look forward, not backward: We must determine if the student is ready for future challenges in which they must transfer prior learning. We should look at whether the student can draw creatively and effectively on their repertoire when handling a novel challenge, not merely determine whether they learned stuff (Wiggins 2011).

The rest of the chapter will guide the reader through strategies and mindsets that can be used by teachers to reclaim assessments *for learning* (Gipps 1994) while also servicing the goal of building inclusive classrooms and assessments. As teachers of large classes ourselves, we are very aware that the usefulness of the strategies outlined here will depend on your current workload, your institution's willingness to support pedagogical change (through resources, funds, cover and support of risk-taking and revolution) and your local teaching and learning climate. Nevertheless, for each strategy and suggestion, we will carefully point out the adoption challenges and opportunities, particularly for larger classes.

5.2 General strategies

If one accepts the premise that all assessments should be formative, then the next step is to reimagine and transform existing summative traditional exams to assessments that promote student learning. The *up-front* cost of transforming assessments to improve learning is necessary and unavoidable, however the burden can be eased by using and developing open education resources (OER) and sharing them between teachers, and even between institutions. That leaves the problem of the *on-going* load increase from administering additional course activities. There is no easy solution, but this is part of the *craft* of teaching: optimizing workflows, effectively managing the course teaching team, leveraging learning technologies and, most importantly, reflecting on and evaluating one's progress towards achieving the pedagogical goals set out in the design of the course. Below are some general philosophies, strategies and pedagogies that underpin the creation of authentic and inclusive assessments.

5.2.1 Frequent testing

Researchers in education and learning have known since at least the early 1930s that frequent assessments in the classroom result in improved learning and retention (Fitch *et al* 1951), particularly with retrieval practice and distributed practice (Benjamin and Tullis 2010). Distributed practice is the act of partitioning one's learning over multiple sessions while retrieval practice involves repeatedly, actively and deliberately recalling information (Gagnon and Cormier 2019) to improve long-term learning and retention. More recent literature has shown demonstrably that infrequent high-stakes assessments (weightings over 20%) such as midterms and exams increase stress and anxiety, favour students that have sustained privilege from higher socioeconomic backgrounds and make our classrooms less inclusive and less diverse (Ballen *et al* 2017b, Cotner *et al* 2017, Salehi *et al* 2020). It should come as no surprise that high-stakes assessments such as midterms and final exams induce extreme test anxiety in students (Harris *et al* 2019) given the consequences and stakes of all that is involved in higher education.

This is usually countered by having more frequent assessments in the form of frequent timed tests, each worth just a few percent of the final grade. However, reducing assessment weight is not enough, as students need to be given opportunities to make up for poor performance on tests or to practice on additional test-like problems. One approach is to implement a 'test/bonus test' structure where students can repeat their test a week later with a completely different set of randomized problems assessing the same concepts. To incentivize students to take these assessments, teachers can take the better score of each 'test/bonus test' pair so students are motivated to attempt the second assessment regardless of their score on the first one. Effectively, this strategy combines both distributed and retrieval practice by having students recall information over multiple sessions in test situations. Even ignoring the learning benefits of distributed retrieval practice, there are also benefits to student engagement with the course, and the classroom community. A major challenge of implementing such a scheme is the need for large problem banks to ensure sufficient question diversity between tests and bonus tests.

For quantitative science disciplines like physics, investing in an algorithmically randomized problem bank can make short work of creating nearly unlimited variants of questions. Each semester, existing questions can be improved and new questions can be added to the bank making it more robust and versatile over time. Sharing banks with other teachers under an open and permissive license can drastically reduce the individual effort and responsibility to maintain the question bank over time. For years, teachers had little choice but to adopt publisher textbooks and offload the costs to students. The rise of publisher-based systems to support digital learning (and digital assessments) have further added additional costs to students, often 'bundling' these costs with textbooks, making it harder for students to use more affordable alternatives like second hand or older editions with minimal differences in content. Now however, with a surge of interest (and investments to support) open educational resources, there are genuine alternatives, particularly for large, introductory science courses. For instance, OpenStax has 50 such course texts, across all domains (Stafford and Flatley 2018). At the University of British Columbia in Canada, promotion of and support for the uptake and creation of OERs has grown steadily over the last decade, now passing direct savings on to students of the order of C$2M per year (see the open.ubc.ca snapshot for details). With randomized problem banks, teachers can provide many opportunities for students to improve their understanding, earn higher scores, meet students wherever they are at, all while retaining the spirit of formative assessments (continued feedback for improvement and growth).

Despite tacit acknowledgement of the success of frequent testing from many teachers and administrators, it is often argued that it is not feasible or practical for teachers to manually create, administer, and grade so many assessments in a semester. It is true that frequent assessments usually come at *some* cost, either by increasing the burden on students with tests done outside of class time, or by reducing the amount of content covered to make room for frequent tests during class. Creative strategies may be devised to strike a different balance between the two extremes such as asking students to learn some material on their own or having the tests during class, and optional bonus tests outside of class. However, personal experience has shown that these strategies usually come with undesired compromises like lower attendance for bonus tests or reduced student satisfaction and motivation. A concrete example of why content reduction **does not** result in lower performance in subsequent courses comes out of recent work from a large, public research university in the United States. Dewsbury *et al* showed that adopting learning-centred pedagogies required cutting about 35% of content from an introductory biology sequence, but the students who were taught less content performed *equally as well or better* in upper-division courses compared to students in a traditionally taught class with more content (Dewsbury *et al* 2022).

In most large classes it is not feasible to conduct frequent tests on paper and learning technologies are needed to scale down the effort required to grade and give feedback to students. Of course, not all authentic assessments can be transformed to a digital format, so some compromises may be needed to balance the degree of authenticity with practicality. But if it is feasible to administer assessments digitally, the possibilities become endless—particularly when learning technologies are leveraged.

5.2.2 Leveraging learning technologies

With increasingly more students pursuing higher education degrees, there is pressure on institutions (and consequently, teachers) to adopt creative solutions to teach more students with the same, or fewer resources. High teaching loads, more non-teaching responsibilities, and an increased expectation for teachers to implement the latest curricular and pedagogical innovations have increased the overall workload of teachers everywhere. Despite these expectations, it has become increasingly difficult for teachers to maintain pedagogical integrity in their classes. Pedagogical integrity is the notion of teachers teaching their classes in ways that are pedagogically sound and supported by evidence-based literature. Teachers often know very well *what they have to do* to promote learning (Schinske and Tanner 2014) in their classes (active learning (Freeman *et al* 2014), frequent assessments (Fitch *et al* 1951), community building (Seidel *et al* 2015), quality timely feedback (Poulos and Mahony 2008), inclusive teaching (Sathy and Hogan 2022)) and the challenge is usually to figure out *how to do it* given limited resources and minimal preparation time while maintaining a healthy work–life balance. Maintaining pedagogical integrity is often practically unattainable due to logistical challenges of scale and the enormous human effort required to achieve the ideal in real classes. Though it can definitely be argued that learning technologies have actually added to this effort burden, we believe the use of learning technologies has now become essential in managing large classrooms at scale.

On their own, learning technologies cannot improve teaching, or the student experience. Buoyed by a strong pedagogical underpinning, teachers can deploy learning technologies to make short work of mundane and menial teaching tasks, usually surrounding assessments. This frees up teachers to appreciate the humanity of instructor–student interactions, develop and evaluate innovative lessons and pedagogies tailored to students in their classes. One major pain point is the inability of teachers to create and maintain effective formative assessments with high-quality automated feedback mechanisms. Another obstacle is the commercial nature of closed-source online assessment platforms, which have exorbitant costs that make adoption in many institutions prohibitively expensive. Open source projects like Jupyter and Quarto provide programming language agnostic ways of making scientific computing accessible to all disciplines and assessments created on these infrastructures can be highly customized and personalized with helper projects like nbgrader and ottergrader. Full service open source platforms like PrairieLearn and Tao Testing allow very sophisticated algorithmic randomization of questions on homework assignments and tests. Where manual grading is necessary, commercial tools like Crowdmark and Gradescope make it much easier to coordinate marking with multiple graders and give students high-quality feedback extremely efficiently, albeit at extremely high costs. Learning analytics tools such as OnTask can be used to send personalized feedback messages regularly in large classrooms. There are even solutions available for traditional paper-based assessments such as AMC (free and open source), PLOM (free and open source), IF-AT cards, and Remark that make manual grading of paper assessments feasible in large classes. If it is truly unfeasible to conduct frequent tests in your classroom or if learning technologies are not accessible or available, we suggest using class time to experiment with interventions that have been shown to mitigate the effects of test anxiety and improve performance on high-stakes exams, *for all students* (Harris *et al* 2019). Some of these strategies will be discussed in the next section.

5.2.3 Transforming assessments with inclusive policies and pedagogies

Though assessment design is a substantive component of inclusive teaching strategies, assessments are secondary to the considerations of safe and equitable learning environments for all students. Assessments are emphasized so much in inclusive classroom conversations solely because as a society, our systems and structures are *outcome-driven* rather than *process-driven*. Without regard to building an inclusive classroom, learning suffers, inequities grow and the student experience is poor. There is a significant amount of work currently being done on the creation of inclusive classrooms, particularly in STEM classes (Harris *et al* 2012). One approach that has shown significant promise (and which is discussed in detail throughout this book) is the adoption of active learning techniques. There is mounting evidence that switching to active learning techniques eliminates performance gaps in equity-seeking groups and actually does not disadvantage *any* group (Ballen *et al* 2017a) through the use of mixed assessment methods and a combination of formative and summative assessment techniques, both low- and high-stakes (Cotner *et al* 2017). Work from the University of Minnesota has shown that mixed assessment methods make biology classes more equitable and reduces performance gaps between male and female students (Ballen *et al* 2017b).

Not only are performance gaps highly context-dependent (Harris *et al* 2019), so too is the success of any intervention used to address them. As Harris *et al* (2019) suggest, before teachers implement any interventions to address persistent performance gaps, it is generally recommended to spend time thinking and interpreting the collected data carefully. Emotional reappraisal is an intervention approach that redirects students to perceive the heightened feelings experienced during stressful situations as helpful for the thinking process (Jamieson *et al* 2016). Expressive writing is another intervention that targets the 'worry' component of anxiety by asking students to write prose about their emotional state before taking exams (Klein and Boals 2001, Harris *et al* 2019). The goal is to 'clear' the working memory of thoughts that are not beneficial to improve test performance (Park *et al* 2014). Other concrete strategies also exist to mitigate performance gaps in various groups due to test design biases (including question type and complexity), stereotype threat and fixed mindsets but their efficacy is mixed and highly contextual (Harris *et al* 2019). Though specific interventions may have variable effects, implementing inclusive teaching pedagogies that create a culture of equity and inclusivity in the classroom reduce outcome gaps and improve long-term performance (Dewsbury *et al* 2022). Sathy and Hogan (2022) have written an excellent book titled, *Inclusive Teaching Strategies for Promoting Equity in the College Classroom* which contains hundreds of strategies (large and small) on how *all* students in your class can feel welcome and included (Sathy and Hogan 2022). Importantly, a framework is provided on how to structure conversations around inclusive teaching, centred around the critical question: 'who is being left behind and what can teachers do to add more structure?'. Often when teachers begin work on making their teaching more inclusive, they start with reflecting on the challenges students in their own classrooms face. It is here where stories about food insecurity, unstable housing, financial concerns, broken homes and a plethora of other issues surface. This is a difficult time for teachers and invariably there is a moment where they ask themselves 'what are we even doing here', and 'does what we do even matter'? When teachers get to this watershed moment, we advocate for embodying a *pedagogy of care*. Pedagogy of care (Palahicky *et al* 2021) is a method and practice of teaching where the instructor takes the role of care-giver and the student takes the role of care-receiver. Maha Bali (2015), a Professor at the American University of Cairo writes:

> Sometimes, the most valuable thing we can offer our students is genuine care for them, their well-being, their happiness. Not just their grades. Not just their learning. But their whole selves.

Bali continues by comparing the medical ethic of 'doing no harm' with the 'educational imperative'[3], both of which often forget to look at the whole human or the overall well-being of the person. Some questions then arise: is our focus on teaching content for literacy and numeracy rather than the overall well-being of our students acceptable? Is doing no harm and teaching mandated curriculum enough for our students? Ideally, most teachers would answer in the negative (putting aside the complexities of feasibility, workload, energy, etc). But practically, how can pedagogies of care be adopted in a classroom? For that, we need to completely reimagine the role of both instructors and students in a classroom. Further we need to critically examine the role of something that has wormed its way into every conversation in higher education over the last two centuries: grades.

5.2.4 Alternative grading paradigms

In *The Trouble with Rubrics*, Alfie Kohn writes: 'Research shows three reliable effects when students are graded: They tend to think less deeply, avoid taking risks, and lose interest in the learning itself.' (Kohn 2006) Most teachers want their students to focus on learning instead of their grades. However, our systems, structures and policies in higher education are heavily centred around grades and the fallacies of their inherent fairness (Schinske and Tanner 2014). In these settings, it is hard to expect students to keep their attention on the material and not get distracted by the frequent input of various grades.

Classical models of teaching focus on differences in student ability when all students are given the same time to learn the same content. An alternative to this philosophy is to flip convention, and instead focus instructional efforts on students

---

[3] The educational imperative is 'To be prepared for today's workforce, informed about important issues, and able to understand the complex world in which we live, all Americans must have a solid education in science, mathematics, and technology' (National Academy of Sciences, National Academy of Engineering, Institute of Medicine 1997).

eventually achieving mastery[4]. Naturally, the progression time for each student will be variable under this model, but old content should generally be mastered before moving on to new material. This educational philosophy is known as mastery learning, first introduced by Benjamin Bloom in 1968 (Bloom 1968). In his paper, Bloom suggested that traditional grading systems focused too much on students' aptitude (how quickly they learn the material) and not enough on their achievements. Bloom suggested a new approach where frequent formative tests would be used to provide feedback and more opportunities for students to prove mastery of a topic. Today, mastery learning is implemented in many courses and contexts, in a variety of different formats but all centred around giving students more flexibility in their learning and multiple attempts to demonstrate mastery. Mastery grading (or mastery-based grading) is the specific assessment approach (rather than the philosophy).

Alternatives to traditional grading systems, such as contract-grading, standards or competency-based grading, specifications-based grading and ungrading, give instructors the freedom to change the conversation and redirect the focus back towards learning. For further discussion on alternative grading systems, consult the recent book written by Clark and Talbert (2023). Suffice it to say, the systems generally included under the alternative grading umbrella are not new, but to varying degrees, all reject the association of learning with points and grades from single attempt assessments. Though these ideas are not new, in recent years, there has been a surge in conversations and discussions around alternative grading paradigms. The precise implementation details are beyond the scope of this chapter, but we invite readers to think outside the box and explore alternative grading paradigms if they resonate with you!

5.3 Specific implementations

The general strategies and philosophies presented in the previous section may be helpful in framing conversations around reimagining assessments for learning. But it may not be possible for a teacher to *only* read descriptions of strategies and learn enough to implement them in their own classrooms. In this section we will provide several practical examples and techniques to transform 'summative assessments' to still have formative properties, so the focus is shifted to designing assessments for learning. Perhaps the most comprehensive resource for creating and using authentic assessments is Jon Mueller's work developing the authentic assessment toolbox (Mueller 2005). We acknowledge that the implementations described below may not be suitable for all contexts and environments, but we hope that it spurs the imagination to consider and adopt alternative assessments that are authentic. We have added references to both the theoretical underpinnings and motivations behind the specific strategies listed below, as well as references to works of scholarship of teaching and learning that demonstrate efficacy (and much more logistical and methodological details) in actual classrooms.

---

[4] We recognize efforts are underway to retire the word 'mastery' (see Talbert (2021) for some discussion on why this term is harmful) but the community has not yet converged on a suitable replacement.

5.3.1 Collaborative two-stage final exams

In collaborative two-stage exams, students first write a test individually and then immediately complete the same (or a very similar) test in groups of three or four students (Gilley and Clarkston 2014). Once the students turn in the individual version of their test, they are free to discuss the questions and answers within their group, and both the individual test and the group test can be done in as little as 50 min (though the experience is smoother with 80 min). There are some policies that can be enacted to minimize the effect of bad actors in a classroom, and maximize the learning benefit for all. For instance, teachers can set the group test as identical to the individual portion, or ask deeper and more challenging questions to leverage the increased expertise of the group. It is also at the teacher's discretion to control how the overall assessment score is distributed between the individual and the group portion. Experience suggests weighting the individual test as 85% of the overall score, and the remaining amount to the group score while instructing students that their score cannot decrease after doing the group test.

Education researchers at the University of British Columbia conducted a rigorous cross-over study to compare the benefits of individual testing versus collaborative testing and determined that the learning benefit more than makes up for the additional time required to administer the two-stage exam. Advocates of two-stage exams frequently remark on the amazing burst of energy that erupts in a room once the group portion starts. Students feel that discussing their questions and answers immediately after doing the test allows them to discuss with their peers while they still care about the content, and while they can still improve their overall score by working in the group. Nearly every student is engaged, feedback is direct and immediate, and learning occurs in a reflective, and collaborative way (Wieman *et al* 2014). Implementing group exams also creates community and increases a sense of belongingness within a classroom (Rieger and Heiner 2014) leading to a more inclusive classroom. However, personal experience has recently revealed that two-stage exams may be a source of acute anxiety in students that have challenges interacting with others in social settings with grades at stake. Thus it is recommended that teachers avoid introducing this pedagogical technique without adequately creating a safe and inclusive space for students to work in. Some examples of conducting two-stage exams more inclusively include simulating the experience with a low-stakes class worksheet prior to the actual exam, making it clear the group portion is optional (and providing an individual post-exam reflection as an alternative), allowing students to choose their own groups (though there are equity and access considerations in group formation), and discussing the pedagogical and learning benefits of the two-stage exam with students to generate and maintain buy-in before and after the two-stage exam. More logistical considerations and examples of two-stage exams are available from Wieman *et al* (2014).

5.3.2 Reflections on exams

In the fast-paced, information-overloaded environments that are typical for learners in higher education institutions, students move through content, courses and programs at break-neck speeds. It is therefore very difficult for students to take stock of past work and truly reflect on it because there's always some other deadline, checkpoint or milestone they need to meet. Through direct intervention, even novice students can be taught to reflect in deep and transformative ways (Ryan 2013). Reflection can be extremely effective for students to crystallize their understanding, course-correct on any misconceptions they have, and take the opportunity to fill in any gaps in their knowledge or understanding. However, students are not naturally proficient at reflecting and self-assessments so appropriate scaffolding must be provided for students to get maximum benefit from reflective learning (Ryan and Ryan 2013). Students often get frustrated with 'qualitative' and 'subjective' questions on assessments—especially if it's a surprise—so it's important to build a culture of reflection throughout the course, rather than just parachute in this style at the end of a course. Some teachers incentivize students to reflect with the opportunity to recover all or a portion of the marks they would have lost on their initial attempt at the assessment if they can effectively demonstrate that they have learned after their initial mistakes.

Unfortunately, there are myriad logistical challenges associated with doing reflection assignments during the final exam period (when traditional summative assessments are held), even if it can be demonstrated these are in service of student learning and well-being. Administrative policies often strictly prohibit teachers from conducting any non-final exam activities during the exam period. In some institutions, there are departmental, faculty and university regulations governing how these examinations are conducted—including minimum and maximum final exam weights (University of Toronto), mandated sit-down scheduled exams for 100- and 200-level courses (UBC), and even mandated randomization versions for large classes (McGill). Currently these (and other) logistical challenges may prevent teachers from asking students to reflect on their final exams, contributing to the culture of learning stopping as soon as the exam starts. However, many of these policies and procedures are remnants and relics from the old days and would benefit from modernization through the lens of inclusivity and a growth-mindset approach towards learning (Dweck 2006). While we wait for these policies and prohibitions to subside, there are some half-measures that are still within the spirit of giving students opportunities to reflect, just without any personalization or feedback, and no direct incentives for the student to actually perform the reflection. One particularly effective strategy is to release 'exemplars', or examples of anonymised students' work of different qualities so students can compare their submissions with other submissions annotated with expert feedback (Scoles *et al* 2013). Another strategy is to incorporate reflective prose-questions on final exams that incentivize students to think deeply about the narrative arc of the course content, and synthesize concepts together. During the COVID-19 pandemic, Francis Su (a Professor of mathematics at Harvey Mudd College in the United States) explored the use of reflective questions to explicitly assess the development of mathematical virtues (mathematical persistence, curiosity, strategy and imagination). Sample questions include: 'How has your mathematical imagination been enhanced as a result of taking this class?', 'Consider one mathematical idea from the course that you have found beautiful, and explain why it is beautiful to you', and 'For any problems you cannot solve on this exam, suggest a strategy you might try to tackle the problem, and show what happened as a result' (Su 2020). In our experience, implementing reflective questions such as these in computer science and physics classes, providing students with these questions in advance produces deeper, richer answers and eliminates the surprise and anxiety of seeing unconventional questions on standard timed final exams. Reading some of the responses to these questions is also truly a joyful and renewing experience!

5.3.3 Automated feedback during final exams

Digital assessments (also called computer-based assessments) can be as simple as scanned paper-based exams that are graded electronically, with or without optical character recognition or AI-assisted tools. More comprehensive digital assessments involve the presentation of randomized questions to students as well as the automatic grading of hundreds of variants that presented to students on any web-enabled device (West *et al* 2021, Perry *et al* 2022). For better or worse, the COVID-19 pandemic has rapidly accelerated the adoption of digital assessments at higher education institutions. To be clear, there are very important access and accessibility considerations associated with adopting digital assessments too quickly (Perry *et al* 2022). However, an argument can be made that aside from the infrastructure requirements of running digital assessments (which should be borne by the institutions and not the individual instructor or student), the barriers to access for digital assessments are no worse than those present with paper exams. Though digital assessments are not quite a panacea, they can mitigate many of the logistical challenges of administering assessments in large classes, particularly when institutions develop computer-based testing facilities (Zilles *et al* 2019). The detailed use of computer-based assessments is discussed in chapter 6 of this book.

Through strategic use of randomized problem banks students can even be given automatic feedback and hints based on their submissions to promote mastery, even within a single assessment session. The natural reaction most teachers will have at the prospect of giving students immediate feedback on a term-end assessment is likely to be dismay and horror. But researchers from the University of Illinois showed that the vast majority of students chose to get immediate feedback on term-end assessments rather than deferring the feedback when given the choice (Verma *et al* 2020). With online assessment platforms, if students are completely stuck, they can even be permitted to request hints for partial credit on questions. Overall, systems that can deliver automated feedback to students are extremely powerful tools that can result in paradigm shifts for the role of assessments in education. As mentioned, there remain challenges in access and accessibility, but these obstacles are by no means insurmountable (Perry *et al* 2022).

5.3.4 Oral examinations

Oral examinations are most simply defined as assessments in which a student's response to a question (or task) is verbal (rather than written) and are an excellent way to more authentically assess students' understanding and learning. They are deeply embedded in several professions including medicine, pharmacy, law and architecture (Joughin 1998). Though oral exams *can* be done as simply as asking students to solve problems that would normally appear on written exams in front of the instructor, this is not the appropriate way to conduct oral examinations and frankly, is a disservice to the rich information that can be gleaned from them (Theobold 2021). The dynamic nature of oral exams allows teachers the flexibility to adjust the order, depth, breadth and type of questions asked on the fly, depending on the responses of the individual student. The primary goal of the oral interaction is to help the student understand their own learning, give them immediate feedback (including praise) on their progress, and ultimately, advance their learning. One should also challenge the student and gradually ratchet up the difficulty as they progress through the oral exam (Westhoff and Hagemeister 2014).

Oral exams often bring about feelings of crippling anxiety and stress amongst students. However, recent work in computer science (Sabin *et al* 2021), mathematics (Nelson 2010), biology (Huxham *et al* 2012) and data science (Theobold 2021) show that student attitudes about oral exams shift drastically after going through the experience, particularly if a low (or no)-stakes practice version is done beforehand. To students the mere *idea* of an oral examination is often more terrifying than the actual reality of the experience. When considering oral exams, the most important thing a teacher can do to address this fear is to give students opportunities to practice the skills of an oral exam in low-stakes environments such as group problem solving sessions, mock oral exams, reflecting on recorded videos of them explaining a concept (Explainer Videos).

Perhaps the only outstanding issue is how to do these exams at scale, in larger classrooms. Frankly, doing oral examinations in classes of over 150 students without a small army of co-instructors and teaching assistants is infeasible unless some compromises are made. For instance, one strategy to scale down the effort needed to administer oral exams is to select a subset of questions to examine students orally. Another strategy is to reduce the time the teaching team spends examining the student by assigning students problem(s) in advance and limiting the teacher–student time just to probing the student understanding, rather than also having students solve a problem they've never seen before (Crannell 1999). Despite the numerous benefits of oral exams and the clear pedagogical advantages, it is not recommended that teachers dive into oral examinations in large classes without adequate resources and support. The University of Wollongong has a comprehensive guide complete with recommendations for planning, executing and evaluating oral assessments (Joughin 2010).

5.3.5 Experiential learning

Of all the options, perhaps the best and most authentic way to assess students is to give them the opportunity to demonstrate their proficiency and understanding through projects and experiential learning opportunities, ideally in team settings. Though it is usually the case that experiential learning opportunities are considered at the *program-level* and offered to cohorts of students in specific and specialized circumstances via capstone projects, co-op options, and work-learn opportunities, these experiences can also be offered to students in individual courses as well. Mazur's introductory physics class at Harvard University is a prime example of a course where students are pushed well outside their comfort zones, take ownership of their own learning, and then demonstrate their progress through two month-long team projects that culminate in a project fair (Mazur and Narang 2021). Project-based introductory physics classes also have the effect of reducing the gender gap in students' self-efficacy in all dimensions studied (Espinosa *et al* 2019) and Mazur's class uses the specification grading system first introduced by Linda Niilson (Nilson 2015). There are no exams at all in Mazur's introductory physics class, nor even any points or grades at all. Students demonstrate their individual knowledge through a variety of assignments, get plenty of feedback from the course team on their progress, and their knowledge is continually assessed through various checkpoints on two term projects. Mazur's goal in this course is to emphasise not just the individual responsibility, but also the team responsibility involved in team projects to make more real-world connections. Other examples of experiential learning in medium to large classes include opportunities to engage with the local community such as Celeste Leander's work integrating salmon farming in her Biol 342 lab ($N = 92$) (Leander 2022), Patrick Culbert's work integrating the forest around the campus into class even when things went online due to the pandemic ($N = 80$) (Culbert 2021), Linda Jennings' work with several biology teachers on incorporating museums in undergraduate teaching ($N = 50 -122$) (Goedhart 2021) and Jonathan Graves' work ($N = 175$) on experiential learning in game theory (Graves 2021). Many (but not all) of these implementations were made possible due to financial support in the form of small institutional grants. These internal grants support innovative teaching transformations by providing one-time financial support for the hiring and training of graduate and undergraduate students to accomplish the work with the intention of the courses becoming sustainable and self-sufficient. Other examples of experiential learning opportunities in a variety of STEM contexts include biology (Hansen *et al* 2021), chemistry (Ginzburg *et al* 2019), data science (Allen 2021), computer science (Sendall *et al* 2019) and engineering (Gadola and Chindamo 2019).

5.4 Conclusion

This chapter was split up into three sections: an introduction which covered the purpose and aspirational goal of assessments, followed by general strategies, philosophies and pedagogies for the creation of authentic assessments, and finally specific implementations to transform traditional summative assessments into more authentic, formative assessments for learning. We've considered several ways to recentre the learner in our classes, implement inclusive teaching policies and develop accessible authentic assessments in support of student learning and well-being. What we have not yet considered is the state of the union on the well-being of teachers everywhere. In short (and in general), teachers are exhausted, and were exhausted even before the COVID-19 pandemic. Teaching workloads are high, administrative and service commitments continue to be ever more burdensome and simultaneously critical to ensure the academy continues to function. It may seem out of touch with reality to write a manuscript about what *more* teachers can do with their limited time and resources. We understand and acknowledge this. We wrote this chapter not to add to the workload of teachers but because we firmly believe that teachers genuinely care about their students and their learning. And so we provided alternatives to the current teaching norms, all in service of institutions to provide additional supports to create and maintain equitable and inclusive learning environments. Students can still continue learning and getting feedback in an authentic way, even on term-end assessments such as final exams. The learning should never stop when *all* assessments are designed to be formative. We hope teachers see the perspectives presented here as light at the end of a dark tunnel—some of it is indeed aspirational and at present, impractical. But much of it *is* feasible with small amounts of effort and large changes in perspective and approach. With enough of us working to change the systems and structures from within, we will move mountains.

References

Archbald D A and Newmann F M 1988 *Beyond Standardized Testing: Assessing Authentic Academic Achievement in the Secondary School* (Reston, VA: National Association of Secondary School Principals Assessment)

Allen G I 2021 Experiential learning in data science: developing an interdisciplinary, client-sponsored capstone program *Proc. 52nd ACM Technical Symp. on Computer Science Education* pp 516–22

Bali M 2015 Pedagogy of care—gone massive *Hybrid Pedagogy* https://hybridpedagogy.org/pedagogy-of-care-gone-massive/

Ballen C J, Wieman C, Salehi S, Searle J B and Zamudio K R 2017a Enhancing diversity in undergraduate science: self-efficacy drives performance gains with active learning *CBE—Life Sci. Educ.* **16** ar56

Ballen C J, Salehi S and Cotner S 2017b Exams disadvantage women in introductory biology *PLoS One* **12** 1–14

Benjamin A S and Tullis J 2010 What makes distributed practice effective? *Cogn. Psychol.* **61** 228–47

Black P and Wiliam D 1998 Assessment and classroom learning *Assess. Educ.: Princ., Policy Pract.* **5** 7–74

Black P and Wiliam D 2004 The formative purpose: assessment must first promote learning *Yearb. Natl. Soc. Study of Educ.* **103** 20–50

Bloom B S 1968 Learning for mastery. Instruction and curriculum. Regional education laboratory for the carolinas and Virginia, topical papers and reprints, number 1 *Eval. Comment* **1** n2

Broadfoot P and Black P 2004 Redefining assessment? The first ten years of assessment in education *Assess. Educ.: Princ. Policy Pract.* **11** 7–26

Clark D and Talbert R 2023 *Grading for Growth: A Guide to Alternative Grading Practices that Promote Authentic Learning and Student Engagement in Higher Education* 1st edn (New York: Routledge)

Cotner S and Ballen C J 2017 Can mixed assessment methods make biology classes more equitable *PLoS One* **12** 1–11

Crannell A 1999 Collaborative oral take-home exams *Assessment Practices in Undergraduate Mathematics* vol 49 ed B Gold, S Z Keith and W A Marion (Washington DC: The Mathematical Association of America) pp 143–5

Crooks T J 1988 The impact of classroom evaluation practices on students *Rev. Educ. Res.* **58** 438–81

Culbert P D 2021 COVID-19 field instruction: bringing the forests of British Columbia to students 8,000 km away *Nat. Sci. Educ.* **50** e20040

Dewsbury B M, Swanson H J, Moseman-Valtierra S and Caulkins J 2022 Inclusive and active pedagogies reduce academic outcome gaps and improve long-term performance *PLoS One* **17** 1–13

Duckworth A L, Quinn P D, Lynam D R, Loeber R and Stouthamer-Loeber M 2011 Role of test motivation in intelligence testing *Proc. Natl, Acad. Sci.* **108** 7716–20

Dweck C S 2006 *Mindset: The New Psychology of Success* (New York: Random House)

Espinosa T, Miller K, Araujo I and Mazur E 2019 Reducing the gender gap in students' physics self-efficacy in a team-and project-based introductory physics class *Phys. Rev. Phys. Educ. Res.* **15** 010132

Fitch M L, Drucker A J and Norton J A Jr 1951 Frequent testing as a motivating factor in large lecture classes *J. Educ. Psychol.* **42** 1–20

Freeman S, Eddy S L, McDonough M, Smith M K, Okoroafor N, Jordt H and Wenderoth M P 2014 Active learning increases student performance in science, engineering, and mathematics *Proc. Natl. Acad. Sci.* **111** 8410–5

Gadola M and Chindamo D 2019 Experiential learning in engineering education: the role of student design competitions and a case study *Int. J. Mech. Eng. Educ.* **47** 3–22

Gagnon M and Cormier S 2019 Retrieval practice and distributed practice: the case of french canadian students *Can. J. Sch. Psychol.* **34** 83–97

Gilley B H and Clarkston B 2014 Collaborative testing: evidence of learning in a controlled in-class study of undergraduate students *J. Coll. Sci. Teach.* **43** 83–91

Ginzburg A L, Check C E, Hovekamp D P, Sillin A N, Brett J, Eshelman H and Hutchison J E 2019 Experiential learning to promote systems thinking in chemistry: evaluating and designing sustainable products in a polymer immersion lab *J. Chem. Educ.* **96** 2863–71

Gipps C 1994 *Beyond Testing: Towards a Theory of Educational Assessment* (London: Falmer Press)

Goedhart C 2021 Using the beaty biodiversity museum collections to engage undergraduate students in science *Bio News: Teach. Spotlight* https://blogs.ubc.ca/bionews/2021/04/27/teaching-spotlight-using-the-beaty-biodiversity-museum-collections-to-engage-undergraduate-students-in-science/ (accessed 20 August 2022)

Graves J 2021 Project AXLRD: experiential learning in game theory *Jonathan Graves: Personal Website* https://jonathanlgraves.arts.ubc.ca/project-axlrd-experiential-learning-in-game-theory/ (accessed 20 August 2022)

Hansen A K *et al* 2021 Biology beyond the classroom: experiential learning through authentic research, design, and community engagement *Integr. Comp. Biol.* **61** 926–33

Harris C, Mullally M and Thomson R 2012 Science is for everyone: integrating equity, diversity, and inclusion in teaching: a toolkit for instructors (Carleton University)

Harris R B, Grunspan D Z, Pelch M A, Fernandes G, Ramirez G and Freeman S 2019 Can test anxiety interventions alleviate a gender gap in an undergraduate STEM course? *CBE—Life Sci. Educ.* **18** ar35

Hogan K A and Sathy V 2022 *Inclusive Teaching: Strategies for Promoting Equity in the College Classroom* (Morganstown, WV: West Virginia University Press)

Huang S, Deshpande A, Yeo S-C, Lo J C, Chee M W L and Gooley J J 2016 Sleep restriction impairs vocabulary learning when adolescents cram for exams: the need for sleep study *Sleep* **39** 1681–90

Huxham M, Campbell F and Westwood J 2012 Oral versus written assessments: a test of student performance and attitudes *Assess. Eval. High. Educ.* **37** 125–36

Jamieson J P, Peters B J, Greenwood E J and Altose A J 2016 Reappraising stress arousal improves performance and reduces evaluation anxiety in classroom exam situations *Soc. Psychol. Personal. Sci.* **7** 579–87

Jenkins L 2013 Permission to forget *J. Qual. Part.* **36** 21

Joughin G 2010 *A Short Guide to Oral Assessment* (Leeds: Leeds Met Press in Association with University of Wollongong) https://eprints.leedsbeckett.ac.uk/id/eprint/2804/

Joughin G 1998 Dimensions of oral assessment *Assess. Eval. High. Educ.* **23** 367–78

Klein K and Boals A 2001 Expressive writing can increase working memory capacity *J. Exp. Psychol.: General* **130** 520

Kohn A 2006 The trouble with rubrics *English J.* **95** 12–5

Kohn A and Blum S D 2020 *Ungrading: Why Rating Students Undermines Learning (and What to Do Instead)* (Morgantown, WV: West Virginia University Press)

Leander C 2022 Why I ungrade, and a how-to primer *Celeste Leander: Pers. Website* https://blogs.ubc.ca/celesteleander/2022/03/17/why-i-ungrade-and-a-how-to-primer/ (accessed 20 August 2022)

Leathwood C 2005 Assessment policy and practice in higher education: purpose, standards and equity *Assess. Evaluat. High. Educ.* **30** 307–24

Lund J 1997 Authentic assessment: its development & applications *J. Phys. Educ., Recreat. Dance* **68** 25–8

Mazur E 2021 Step into the classroom of Eric Mazur instructional moves: (Harvard Initiative for Learning & Teaching) https://instructionalmoves.gse.harvard.edu/eric-mazur#moves (accessed 5 July 2023)

Mueller J 2005 The authentic assessment toolbox: enhancing student learning through online faculty development *J. Online Learn. Teach.* **1** 1–7

National Academy of Sciences, National Academy of Engineering, Institute of Medicine 1997 *Preparing for the 21st Century: The Education Imperative* (Washington, DC: The National Academies Press) https://nap.nationalacademies.org/catalog/9537/preparing-for-the-21st-century-the-education-imperative

Nelson M A 2010 Oral assessments: improving retention, grades, and understanding *PRIMUS* **21** 47–61

Nilson L B 2015 *Specifications Grading: Restoring Rigor, Motivating Students, and Saving Faculty Time* (Sterling, VA: Stylus Publishing LLC)

Palahicky S, DesBiens D, Jeffery K and Webster K S 2021 Pedagogical values in online and blended learning environments in higher education *Research Anthology on Developing Effective Online Learning Courses* (Hershey, PA: IGI Global) pp 1316–38

Park D, Ramirez G and Beilock S L 2014 The role of expressive writing in math anxiety *J. Exp. Psychol. Appl.* **20** 103

Perry K, Meissel K and Hill M F 2022 Rebooting assessment: exploring the challenges and benefits of shifting from pen-and-paper to computer in summative assessment *Educ. Res. Rev.* **36** 100451

Poulos A and Mahony M J 2008 Effectiveness of feedback: the students' perspective *Assess. Eval. High. Educ.* **33** 143–54

Rieger G W and Heiner C E 2014 Examinations that support collaborative learning: the students' perspective *J. Coll. Sci. Teach.* **43** 41–7

Roediger H L and Karpicke J D 2006 Test-enhanced learning: taking memory tests improves long-term retention *Psychol. Sci.* **17** 249–55

Rosales J and Walker T 2018 The racist beginnings of standardized testing *Natl. Educ. Assoc.* **73** 88–96

Ryan M 2013 The pedagogical balancing act: teaching reflection in higher education *Teach. High. Educ.* **18** 144–55

Ryan M and Ryan M 2013 Theorising a model for teaching and assessing reflective learning in higher education *High. Educ. Res. Devel.* **32** 244–57

Sabin M, Jin K H and Smith A 2021 Oral exams in shift to remote learning *Proc. 52nd ACM Technical Symp. on Computer Science Education* pp 666–72

Salehi S, Cotner S and Ballen C J 2020 Variation in incoming academic preparation: consequences for minority and first-generation students *Front. Educ.* **5** 552364

Schinske J and Tanner K 2014 Teaching more by grading less (or differently) *CBE—Life Sci. Educ.* **13** 159–66

Scoles J, Huxham M and McArthur J 2013 No longer exempt from good practice: using exemplars to close the feedback gap for exams *Assess. Eval. High. Educ.* **38** 631–45

Seidel S B, Reggi A L, Schinske J N, Burrus L W and Tanner K D 2015 Beyond the biology: a systematic investigation of noncontent instructor talk in an introductory biology course *CBE—Life Sci. Educ.* **14** ar43

Sendall P, Stuetzle C S, Kissel Z A and Hameed T 2019 Experiential learning in the technology disciplines *Proc. EDSIG Conf.* vol 2473 p 4901

Stafford D and Flatley R 2018 OpenStax *Charlest. Advis.* **20** 48–51

Stalnaker J M and Stalnaker R C 1934 Open-book examinations *J. High. Educ.* **5** 117–20

Stray C 2001 The shift from oral to written examination: Cambridge and Oxford 1700–1900 *Assess. Educ.: Princ., Policy Pract.* **8** 33–50

Su F 2020 7 Exam questions for a pandemic (or any other time) *Francis Su: Personal Website* https://francissu.com/post/7-exam-questions-for-a-pandemic-or-any-other-time

Talbert C 2021 A word about words *Robert Talbert, Ph.D.: Personal Website* https://rtalbert.org/a-word-about-words/

Theobold A S 2021 Oral exams: a more meaningful assessment of students' understanding *J. Stat. Data Sci. Educ.* **29** 156–9

Verma A, Bretl T, West M and Zilles C 2020 A quantitative analysis of when students choose to grade questions on computerized exams with multiple attempts *Proc. 7th ACM Conf. on Learning@ Scale* 329–32

Viruru R 2006 Postcolonial technologies of power: standardized testing and representing diverse young children *Int. J. Educ. Policy, Res., Pract.* **7** 49–70

Vygotsky L S 1978 *Mind in Society: Development of Higher Psychological Processes ([Ca. 1930–1934])* (Cambridge, MA: Harvard University Press) http://jstor.org/stable/j.ctvjf9vz4

West M, Walters N, Silva M, Bretl T and Zilles C 2021 Integrating diverse learning tools using the PrairieLearn platform *7th SPLICE Workshop at SIGCSE*

Westhoff K and Hagemeister C 2014 Competence-oriented oral examinations: objective and valid *Psychol. Test Assess. Model.* **56** 319

Wieman C E, Rieger G W and Heiner C E 2014 Physics exams that promote collaborative learning *Phys. Teach.* **52** 51–3

Wiggins G 2011 Moving to modern assessments *Phi Delta Kappan* **92** 63

Williams J B and Wong A 2009 The efficacy of final examinations: a comparative study of closed-book, invigilated exams and open-book, open-web exams *Br. J. Educ. Technol.* **40** 227–36

Zilles C B, West M, Herman G L and Bretl T 2019 Every university should have a computer-based testing facility *CSEDU (1)* 414–20

IOP Publishing

Effective Teaching in Large STEM Classes

Anna K Wood

Chapter 6

Computer-marked assessment and concept inventories

Sally Jordan

After discussing the use of computers in assessment more generally, this chapter concentrates on online computer-marked assessment, in which responses are automatically marked and feedback is instantaneously provided to students. Concept inventories, designed to assess students' conceptual understanding, can make use of the same technologies. I consider the advantages and disadvantages that computer-marked assessment can offer and ways in which its quality can be improved by thorough assessment design, choice of appropriate question types, and careful question writing. I conclude that computer-marked assessment is not a panacea, but nevertheless has much to offer to teachers and learners in large STEM classes.

6.1 Introduction

When considering the needs of large classes and their teachers, the use of a computer to mark and deliver feedback to both students and their teachers is an immediately attractive proposition. However, this use of technology requires particularly careful implementation. Chapter 5 illustrated the potential that assessment has to bring learning benefits, but also pointed out some of the risks. There is a danger that even well-meaning assessment and feedback interventions may create a barrier to learning rather than enabling it (Bangert-Drowns *et al* 1991). These risks are greater when the assessment is delivered remotely or marked by electronic means, especially when there is no human intermediary. As Ridgway *et al* (2004, p 7) comment, 'when we consider the introduction of e-assessment we should be aware that we are dealing with a very sharp sword'.

In this chapter I will explore the advantages and disadvantages of computer-marked assessment and concept inventories. I will discuss ways in which the advantages can be reinforced, and the disadvantages ameliorated, by careful assessment design, choice of appropriate question types and careful question writing. Throughout, I will summarize some of the underpinning theory, while also offering practical hints that derive from my own experience.

6.2 Definitions and history

To avoid any confusion later in the chapter, in this section I will define what I mean by 'computer-marked assessment', 'concept inventory', and some related terms. I will also give a brief history of work in this area.

Since the early years of the 21st century, the term 'e-assessment' (electronic assessment) has been used to include any use of a computer as part of any assessment-related activity (JISC 2006). Terms such as 'digital assessment', 'computer-based assessment' and 'technology-enhanced assessment' are similarly broad, though each brings a slightly different emphasis. These terms encompass, among many other things, the use of e-portfolios, the delivery of audio or video feedback, and the online delivery of written assessments to a teacher for marking, and their later return to students. In recent times, much attention has been given to remote online examinations and the detection of plagiarism and contract cheating (Dawson *et al* 2020).

More specifically, computer-marked assessment refers to situations in which students' responses are automatically marked and, in some cases, feedback is automatically generated. The earliest computer-marked multiple-choice questions were probably E L Thorndike's Alpha and Beta tests used to assess recruits for service in the US Army during the First World War; during the 20th century multiple-choice questions also gained in popularity as an educational tool. By the 21st century the focus became online computer-marked assessment, which enables instantaneous interaction between a student and the system on which the online assessment sits. At the same time, the number of providers of computer-marking assessment systems has grown and the range of question types has extended way beyond multiple-choice questions. For a more detailed historical approach, see Jordan (2013).

A particular use of computers in assessment, common in STEM, is in concept inventories. A concept inventory is an instrument designed to assess students' conceptual understanding, usually with the aim of measuring the learning gain that has occurred across a class as a result of a particular piece of teaching (Sands *et al* 2018). Following the practice established in the Force Concept Inventory (FCI) (Hestenes *et al* 1992), believed to have been the first instrument of this type, most current concept inventories consist of a series of multiple-choice questions, each with one correct answer and a number of incorrect answers, known as distractors, based on common student misconceptions. In order to measure learning gain, they are presented to students as a 'pre-test' before the pedagogical intervention and then repeated as a 'post-test'. Although concept inventories are still sometimes run as paper-based exercises, they are increasingly also run online, which makes them quick and easy to administer, even when considering large class sizes. Students are not usually given direct feedback on their answers to concept inventories, but instead feedback is provided to the teacher, thus fulfilling one of the functions that has been identified if assessment is to support learning (Nicol and Macfarlane-Dick 2006). It is therefore logical to consider concept inventories in this chapter alongside other uses of computer-marked assessment in supporting effective teaching in large class sizes.

6.3 Advantages and disadvantages of computer-marked assessment

6.3.1 Why use computer-marked assessment?

Computer-marked assessment brings many affordances. It has the potential to mark and deliver feedback on students' work, while saving academic time and therefore money. Indeed, Boitshwarelo *et al* (2017) see the recent growth in the use of online quizzes as an inevitable corollary of the dual drivers of reduced resources for teaching and growth in student numbers. Online quizzes can be re-used from year to year at minimal cost. Furthermore, provided a student has access to an appropriate device and the internet, computer-marked assessment can be completed from any location, something which became highly relevant during the COVID-19 pandemic.

The phrase 'objective questions', used historically to describe multiple-choice questions, reflects the fact that the early use of computer-marked assessment came from a desire to make assessment more objective. Ashburn (1938) noted a variation in the grading of essays by different markers, and issues with the reliability of human grading of essays (Brown 2010) and short-answer questions (Butcher and Jordan 2010) remain a persistent concern. Human markers are inherently inconsistent (Bloxham *et al* 2016) and they can be influenced by their expectations of individual students. Even for more open-ended questions, computerized marking brings objectivity and a consistency that can never be assured between human markers (inter-rater reliability) or, over time, for the same human marker (intra-rater reliability).

Computer-marked assessment also has the potential to enhance students' learning. Computer-generated feedback can be provided tirelessly and instantaneously, without the delay inevitably caused by a human marker, and students can be given the opportunity to immediately repeat the task or to perform a similar one and so to learn from the feedback provided. Thus, the feedback is received by students 'while it still matters to them and in time for them to pay attention to further learning or to receive further assistance', fulfilling one of Gibbs and Simpson's conditions under which assessment supports learning (Gibbs and Simpson 2005, p 18).

Tests can be offered to students regularly and, in formative use, even the simplest of quizzes can enable students to repeatedly check their own understanding, encouraging self-regulated learning (Nicol and Macfarlane-Dick 2006). Regular computer-marked assessment can also help students to pace their study, and many authors speak of its role in engaging students and motivating learning (e.g., Holmes 2015) and building self-efficacy and confidence (Cassady and Grindley 2005). Riegel and Evans (2021) found that students experienced a range of positive emotions such as hope and pride more strongly when responding to an online mathematics quiz rather than a conventional test, and a range of negative emotions such as anxiety, anger and hopelessness were all rated less strongly. One student commented that 'online quizzes make me relax and it is enjoyable to me' (Riegel and Evans 2021, p 82). In a survey I conducted (Jordan 2011, p 153), 64%–68% of students agreed with the statement that answering quiz questions was 'fun', and comments such as 'It's more like having an online tutorial than doing a test' and 'give[s] you confidence that you're heading on the right lines' were received.

Computer-generated feedback is inevitably impersonal and non-judgemental, and many students appreciate being able to make mistakes in private (Miller 2009). Although the feedback may be relatively detailed and targeted to specific errors in the student's answer, the fact that it is not responding to a specific person inevitably means that the focus is on the student's performance rather than on the student themselves, which is seen as a key feature in enabling the student to learn (Gibbs and Simpson 2005).

Modern conceptions of effective assessment feedback see this as a *process* with student sense-making and action to improve at the centre rather than as a *product* delivered to students (Sadler *et al* 2023, Winstone and Carless 2020). The feedback provided on computer-marked assessment is most effective when regarded as information that is provided to students, for them to make sense of and act on themselves. Although the assessment is automatically marked and feedback is automatically generated, the student remains in control of their own learning.

Many of the claims for the positive impact of computer-based assessment on learning rely on student opinion; this is important, but in addition to improving student satisfaction, we want to improve learning itself. Furthermore, even when there is a demonstrable correlation between some claims of improved course outcomes and engagement with computer-marked assessment, it can be difficult to be sure that the link is causal. For example, in the case of a purely formative online quiz, it is likely that the more diligent students will both engage with the quiz and do better in the course overall; that does not mean that engagement with the quiz led to the better overall result. However, there have been a number of more rigorous studies that have demonstrated a positive influence of computer-based assessment (e.g., Van Gaal and De Ridder 2013). In addition, research into the so-called 'testing effect' shows that the mere act of taking tests leads to an improvement in subsequent performance that is greater than additional study of the material, even when the tests are given without feedback (Roediger and Karpicke 2006, Prisacari 2015).

Most computer-marked assessment systems are also able to provide information to teachers about the performance of individual students and of the class as a whole, both on individual questions and on the assessment overall. At the whole-class level, this enables improvements to be made to the assessments and the feedback provided. It can also inform the teacher's teaching. For example, if a quiz or concept inventory is used with a class before a lecture or other intervention, it can also inform the teacher where they need to concentrate their efforts. If the use of the concept inventory is repeated after the intervention, its effectiveness can be gauged.

In recent years, the growth of data about students and their online behaviour, and the ability to analyse this, has led to the new field of learning analytics, defined by the Society for Learning Analytics Research as 'the measurement, collection, analysis and reporting of data about learners and their contexts, for purposes of understanding and optimizing learning and the environments in which it occurs'[1]. As the prevalence of online teaching and learning increases, there is an increased potential to make use of an individual student's digital footprint, including their use of e-assessment of all types, to target advice and teaching interventions to their particular needs. This approach raises some ethical concerns (Kitto and Knight 2019) but has been seen to markedly increase retention and completion rates (de Oliveira *et al* 2021).

6.3.2 Why not?

For all its advantages, concerns have been expressed about increasing reliance on computer-marked assessment (Or and Chapman 2022). In this section I will discuss the potential disadvantages. Perhaps the most significant of these relate to lack of authenticity. In a scathing comment about the over-use of multiple-choice questions in medical education, Mitchell *et al* (2003, p 252) quotes Veloski (1999): 'Patients do not present with five choices'. Bridgeman (1992, p 271) makes a similar point with reference to engineers and chemists: they are seldom 'confronted with five numerical answers of which one, and only one, will be the correct solution'. Even when more sophisticated question types are used, there is some anxiety that use of computer-marked assessment will encourage a surface approach to learning.

Other criticisms of multiple-choice questions include the fact that students can guess the answer, or work backwards from the distractors. For example, if a question asks students to integrate an expression and offers five possible answers, they do not necessarily have to know how to integrate to work out the correct answer; they could instead differentiate each of the options until they arrive at the original expression (Sangwin 2013). In this case, the question is not actually addressing what it set out to assess.

Even so-called constructed-response question types (questions that require an answer to be entered into the system, in contrast to selected-response questions in which the students pick an option or options from a choice that is made available) usually require the entry of a single answer. Thus, it is the answer that is being marked, not the working, something which is at variance with much assessment in STEM subjects, where students are encouraged to show their working and explain their logic. Various attempts have been made over the years to replicate the way in which humans mark, but I am not aware of any that have fully succeeded. The most common approach has been to break a question down into constituent steps, which brings the advantages of scaffolding for less confident students (Dawkins *et al* 2017),

---

[1] Society for Learning Analytics (SOLAR) www.solaresearch.org

but it does not truly replicate the more open-ended task it was seeking to replace. This leads to various potential problems. If the assessment is summative, it is not possible to give credit for partially correct answers. If the focus is formative, it may not be possible to identify the source of a student's difficulty and so to give appropriate feedback; were they floundering at the start of the problem or did they make a careless slip near the end? Finally, this lack of evidence for the working behind a final answer can, in principle, make it easier for a student to cheat. It is easier for a student to copy a single answer from a fellow student or a 'homework' website. Plagiarism is discussed further in section 6.3.

In section 6.1, I discussed the preference of some students to receive feedback from a computer rather than a person. However, the lack of a human intermediary between the assessment and the marking engine also brings disadvantages. While it is possible for a tutor to be available to explain an ambiguous question to a student, in much the same way as they would for any type of assessment, there is not usually a human available to explain an unexpected response from a student to the computer! This can lead to inaccurate marking, even for multiple-choice questions. More commonly, the fact that feedback is usually pre-prepared in the expectation of the wrong answers that will be given, rather than being given in response to the actual answer received, means that the feedback may fail to respond in a helpful way to the answer given by a particular student.

6.3.3 Disadvantage or advantage?

In considering the disadvantages of computer-marked assessment, which undoubtedly do exist, it is important not to also blame the use of technology for difficulties that have other causes (Bull and Dyson 2004). Many concerns over plagiarism fall into this category. During the COVID-19 pandemic, it became necessary for students to complete assessed tasks from their own home rather than in invigilated examination halls, and this led to an increased incidence of plagiarism (Montenegro-Rueda *et al* 2021). However, the fact that many institutions increased their use of computer-marked assessment in response to the need to assess remotely does not mean that the computer-marked assessment was the cause of the rise in plagiarism, but rather the necessarily remote location from which it was being completed. Indeed, different variants of computer-marked questions can often be produced for minimum effort, as discussed later, meaning that it is relatively easy for different students to receive subtly different assessments, limiting opportunities for plagiarism.

Similarly, the use of computer-based assessment, in common with any educational intervention which relies on access to a computer and the internet, raises some concern over equitable accessibility for all students. If students are completing the assessment on their own device and from their own home, then there are indeed legitimate concerns relating to digital poverty and the digital divide. However, the use of a technology which means that students can complete an assessment without travelling can save transport costs and make the assessment more accessible to those with certain disabilities as well as to those who are unable or reluctant to travel for other reasons, for example caring responsibilities. In addition, most computer-marked assessment systems include features designed to increase accessibility for those with eyesight problems or dyslexia etc. Accessibility is discussed further in section 6.4.

The feedback from students that has given me the most pleasure, relating to the use of computer-marked assessment in the Open University's STEM Faculty, is that which describes positive enjoyment in completing an assessment (even when with a high-stakes summative function) and that which talks about the assessment feeling as if there was a human tutor there to guide their learning. However, the feedback we receive is not uniformly positive. Careful analysis of the negative feedback received, in the light of the questions and feedback it relates to, has led me to conclude that student complaints usually originate as a result of a particular question, most commonly because the wording is ambiguous or the student has been given misleading feedback (Jordan 2011). My own experience of online quizzes in everyday life strengthens my view that whether computer-marked assessment is a 'good thing' or a 'bad thing', and the extent of its effectiveness, depend on the details of its operation, and context, that can all too easily by overlooked. Sections 6.4–6.6 explore these points more fully.

6.4 Assessment design and integration with teaching

In seeking to develop high-quality computer-marked assessment, it is important to start by thinking about why you are taking this approach in the first place, and how it links to your teaching and other assessment. It is also important to think about any limitations imposed and affordances offered by the particular system that you are using, and to remember that the computer-marked assessment needs to be accessible to all students. In this section I will consider these points in turn, in each case building on my own experience.

6.4.1 Motivation and purpose

I used computer-marked assessment for the first time in the Open University (OU) course *Maths for Science* that ran from 2002 to 2018. The course was the first in the University to use online computer-based assessment of the type described in this chapter, though the colleagues on whose shoulders we built had previously sent questions out to our distance-learning students by CD-ROM and DVD, and for many years we had been using multiple-choice questions which students answered on a prepared computer-readable form. In making the decision to move to online computer-based assessment, we had to assume that students had access to a computer and the internet, at home, work or via an 'internet café' or library. We also relied on the University's systems capability to record student attempts at the computer-marked assessment and to transfer information about performance to the student's record. Our decision to go ahead thus assumed a certain level of technological readiness, but the primary reason for the move was pedagogical not technological; I had a very strong wish to provide instantaneous and targeted feedback to our students on their mathematical ability, with an opportunity to repeat the question, and the move to online computer-marked assessment enabled this. Similarly, although students were required to achieve a certain overall mark in order to pass the module, my motivation for introducing computer-marked assessment was for its formative not its summative function.

More generally, before deciding to use any particular type of assessment, it is important to reflect on what your purpose is: summative ('assessment of learning') or formative ('assessment for learning')? Or is your aim to find out what your students already know, or to ascertain their conceptual understanding so as to investigate the effectiveness a teaching intervention? It is also important, though surprisingly uncommon, to think about the learning outcomes that you are seeking to assess. As a general rule, learning outcomes at the lower end of Bloom's taxonomy (Bloom *et al*, 1956) such as those related to recall are perhaps more easily assessed by simple computer-marked questions than higher-order learning outcomes.

6.4.2 The context in which the assessment is used

Individual teachers may or may not be able to influence the overall assessment strategy of a course or a qualification. However, apparently small differences in assessment strategy can make a very large difference to the effectiveness of individual components, including computer-marked assessment. An example of this was the introduction of a formative thresholded assessment strategy across the Open University's Science faculty from the early 2010s. This strategy required students to reach a modest threshold on the course's continuous assessment, but their continuous assessment score did not contribute to their overall course outcome. The detail of and rationale for this strategy is not relevant to this chapter, but the key point is that two separate models were used, one of which simply required students to reach a threshold of just 30% in, say, five out of seven of the interactive computer-marked assignments (iCMAs). In addition to being allowed three tries with increasing feedback, students could also repeat questions or the whole iCMA as many times as they wanted to, with multiple variants available. This model was found to be very successful, with many students repeating the questions in different variants and, apparently in consequence, also doing better on the overall course outcomes.

In addition to its direct use in formative and summative assessment, computer-marked assessment has a role to play in many of the effective teaching techniques mentioned throughout this book, for example in peer instruction (introduced in chapter 2). The function, use and effectiveness are different depending on context. This serves as a useful reminder of the fact that technology is merely a tool that can be useful in supporting teaching, learning and assessment when appropriate. Similarly, any particular type of assessment is simply a tool in the hands of the teacher.

Thus it is that concept inventories, while potentially using the same technology as other types of computer-marked assessment, have a different function, namely to provide feedback to the teacher about their students' understanding at a specific point in the course and so (usually) to measure learning gain. For this reason, and in contrast to most uses of computer-marked assessment, common practice is for no direct feedback on their performance on a concept inventory to be given to students, partly because of concerns about widespread circulation of questions and their answers.

Another way in which computer-marked assessment can be used in effective STEM teaching is by requiring students to author the questions, for example using the PeerWise system, which enables students to create questions for their fellow-students to answer. Students may also be encouraged to provide feedback to question authors, and to engage in discussion about the questions with their peers. A significant positive correlation has been found between engagement with PeerWise and overall attainment on a range of STEM courses, even after controlling for ability prior to the course (Kay *et al* 2020).

6.4.3 How does the assessment run?

When writing computer-marked assessment, it can be tempting to think about question type but not to give much thought to the way in which the questions run within the overall quiz. The growth of research into the impact of gamification in education (Dichev and Dicheva 2017) reminds us of the need to think more broadly.

There is some variation between the functionality of different quiz engines and virtual learning environments, but most now allow a particular instance of a question to be attempted several times, with retry allowed after receipt of a feedback hint. In some systems the feedback can be varied depending both on how many attempts the student has had and on the answer they give. I most commonly use questions that operate in the manner illustrated in figure 6.1. After their first incorrect try (shown in the top left image), students are simply told that their answer is incorrect, to give them an attempt to find and rectify their error for themselves. After a second incorrect try, more detailed feedback is given, where possible targeted to the student misunderstanding. After the third try, whether the answer given is correct or incorrect, a full answer is given. If a numerical score is required, either for summative use or for the purpose of generating feedback on the quiz overall, a decreasing score can be awarded depending on whether the answer is correct, partially correct or incorrect at each of the three tries.

If allowed by the assessment system and the overall assessment strategy, it may also be possible to repeat the whole question, hopefully in a different variant. So, in the simple example shown in figure 6.1, if a student chose to repeat the whole question, they might be asked to find $1/4 + 1/3$. When the focus is formative, some authors will write variants of questions that differ by a greater amount than simply changing the numbers, to provide greater variety. Alternatively, it may be possible to select different questions from a question bank. In summative use, different variants of questions can be used as an anti-plagiarism device, moving towards a situation in which each student receives a different assessment. However, in this case, it is important to ensure that the different variants used assess the same learning outcome and are of similar difficulty. Up to a point, this can be achieved by careful review of the variants, for example recognizing that calculations involving very small numbers (with negative powers of 10 when expressed in scientific notation) tend to be more difficult than those involving very large numbers.

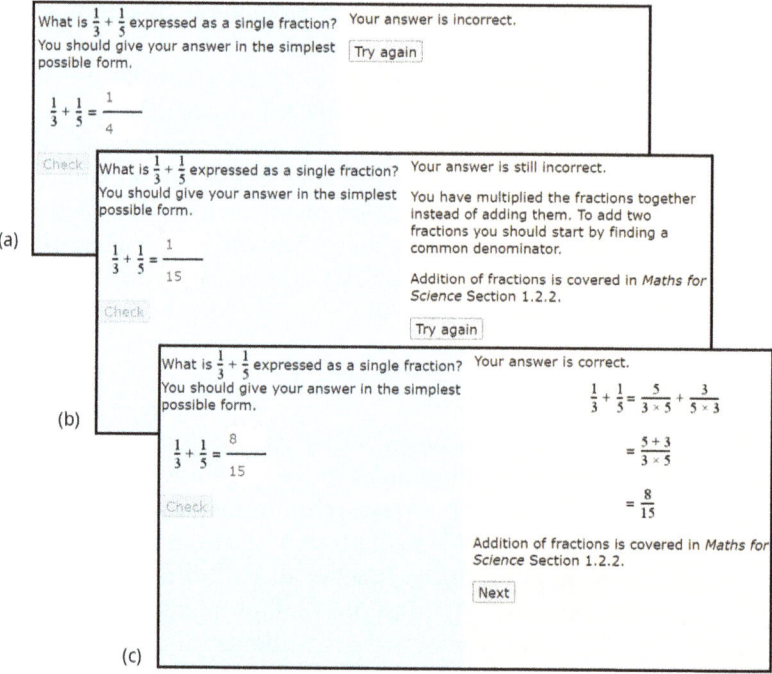

Figure 6.1. A simple question, showing three tries at a question with increasing feedback. This question was written in the OU's OpenMark system, whose functionality informed the development of the Moodle™ Quiz Engine. Screenshot taken from OpenMark. Copyright The Open University. Made available under the GPLv3 license.

Sometimes, variants that are more difficult than others (perhaps because there is some additional skill being assessed) are not spotted until the performance analysis that should be done after the assessment has run. This is discussed further in section 6.7.3.

I have never imposed tight time constraints on quizzes that I have authored, partly because of the context in which I operate: OU students are frequently studying part-time alongside other responsibilities and they are usually studying from home, so interruptions can be difficult to avoid. However, my reluctance to impose a strict time limit goes beyond my own context; the pressure caused by an awareness that time is running out can impair a student's ability to complete the quiz to the best of their ability and can act as a barrier to learning. I have however, imposed hard cut-off dates on quizzes (days, weeks or months from when the quiz was made available), to encourage student pacing through the course. The consideration of time limits and deadlines is yet another matter that depends on the context in which you operate, the purpose of the assessment, and the learning outcomes being assessed. You may, for example, be explicitly aiming to assess a student's ability to work under pressure.

As well as generating feedback on a student's answer to each question, it is usually possible to generate appropriate feedback based on a group of questions assessing the same learning outcome, or on the whole quiz.

6.4.4 Accessibility

As mentioned in section 6.3, the use of computer-marked assessment brings some advantages with regards to accessibility, but there are also some issues that require consideration. Early computer-marked assessment systems relied on internal systems to enable the magnification of text, to alter the text or background colour (which some students with dyslexia find helpful) or to produce a plain text version suitable for feeding into a screen reader. Modern web-based systems more commonly make use of the accessibility systems in the browser that the student is using, which brings alignment with other online tools, but also requires question developers to be more aware of the wider provision available.

Some question types are more difficult to use than others for students with limited dexterity. For example, 'drag and drop' questions, which require students to drag an option into place, require relatively fine motor skills, including the use of a mouse or touchpad. However, from a functional if not an aesthetic perspective, a drop-down list of options can be provided as a substitute for the draggable options, and this list can be navigated by keyboard functions or read by a screen reader.

As for any sort of teaching or assessment resource, figure and graph descriptions should be provided for the use of those who are blind or partially sighted, or who benefit from spoken versions of the questions for any reason. However, it is worth noting that the use of a figure or graph description may result in a change to what the question is assessing. For those who do not need to use a screen reader but whose eyesight is sub-optimal, generous tolerances should be placed on the range of acceptable answers to any questions that require students to read values from a graph etc.

At the OU, there remain a small number of students for whom access to online resources is problematic. This group includes a few students with particular disabilities, e.g., epilepsy, students studying in some prisons and other secure institutions, where access to the internet is not allowed, and—rarely—students who are unable to access the internet because of an unexpected technical problem. We make alternative versions of the questions available in these circumstances, while recognizing that the assessment alters as a result, and the students do not benefit from the instantaneous feedback. Designing assessments to be useable from mobile phones and tablets enables wider accessibility.

Taking a broad definition of 'accessibility', I would also emphasize the importance of minimizing the use of extraneous contextual information in questions. There is a popular belief that this adds interest, which may be the case for a small number of students. However, it more often confuses students, for example if the context is a sport which is unfamiliar to those from different cultural backgrounds.

6.5 Question types

Most computer-marked assessment systems offer a range of question types. In this section I will consider the most common and introduce some less common question types which have particular potential for assessing large STEM classes. Some of the question types introduced here will be discussed in more detail in the case study chapters later in this book.

6.5.1 Selected-response questions

Selected-response questions, defined as those in which a student selects from pre-defined options, are most commonly multiple-choice or multiple-response (in which students are required to select more than one option) but this category also includes true/false questions, questions which require students to match one statement to another, and drag and drop questions. These question types, especially multiple-choice, are generally considered easier and faster to write than constructed-response questions. Care must still be taken in question writing, and the criticisms of lack of authenticity, being able to work backwards, and being able to arrive at the answer by guesswork generally apply to all selected-response questions. However, it becomes more difficult to arrive at the correct answer by guesswork if students are required to select several options.

Other ways of discouraging guesswork include requiring students to explain their answer in a free-text box. This answer is not necessarily marked, but it is available to the teacher should they wish to check that a student understands why an answer is correct. Similarly, students can be asked to upload a file containing their working. Another way of discouraging guesswork is so-called confidence-based (or certainty-based) marking, in which students are required to rate their confidence as well as to give an answer. A correct but unconfident answer receives a lower score than a correct confident answer, whereas an incorrect confident answer is more heavily penalized than an incorrect unconfident one (Gardner-Medwin 2019).

For all the criticisms of them, even simple selected-response questions can lead to 'moments of contingency', formative interactions that can improve cognition (Black and Wiliam 2009). This enables 'catalytic assessment', the use of simple questions to trigger deep learning (Draper 2009). In addition, there are some situations in which a selected-response question is the most suitable type to use. A carefully worded selected-response question, such as the one shown in figure 6.2, can require a certain amount of logical reasoning and thus assess learning outcomes of higher order than simple recall questions.

6.5.2 Simple constructed-response questions

The answer matching required for a question in which the student enters a simple numerical answer should be straightforward, rendering it unnecessary to ask this sort of question in multiple-choice form. It becomes more complicated if you require the answer to be in scientific notation, or to a particular precision. However, many modern quiz engines and virtual learning environments now include specific functionality to enable such questions to be automatically marked.

Similarly, questions that can be answered in the form of one or a small number of letters and other symbols, e.g., 'Give the standard abbreviation for the SI unit of mass' can be automatically marked by relatively straightforward means such as string matching. However, immediate thought must be given to whether the case of

The statements in the following list all refer to the description of motion. Check the boxes of the THREE TRUE statements.

☐ 1. It is possible for a particle to move along a straight line with a positive instantaneous acceleration ($a_x > 0$), and to be slowing down.

☐ 2. When a package is dropped from an aircraft flying horizontally, it hits the ground at a point vertically below its point of release from the aircraft.

☐ 3. If two particles move in uniform circular motion in circles of radii r_1 and r_2 respectively, and each takes the same time to complete one orbit, the particle with the greatest radius of orbit has the greatest magnitude of acceleration.

☐ 4. If a particle undergoes uniform circular motion in a horizontal plane, moving clockwise around a circle as seen from above the plane, the angular velocity vector of the particle points vertically upwards.

☐ 5. In simple harmonic motion, the magnitude of the acceleration of the particle is greatest when the particle is instantaneously at rest.

☐ 6. Each planet moves in a ellipse around the Sun with the Sun at the intersection of the major and minor axes of the ellipse.

Figure 6.2. A multiple-response question.

the letters and order in which they are written is significant: in the example given, kg is a correct answer but KG and gk are not, though I would give targeted feedback for KG. If the correct answer is an algebraic expression, e.g., 'vt', then 'tv' is also correct. Some systems have specific functionality to check for the correct units alongside numerical values.

6.5.3 Questions based on computer-algebra

The introduction of assessment underpinned by computer-algebra systems (CAS) has revolutionized the computer-marked assessment of mathematics and subjects like physics and engineering. The work described in chapter 9 is based on the use of one of the leading and well-established systems, STACK, which is underpinned by Maxima. The system enables students to enter an answer as a mathematical expression, which STACK then asks the student to verify as being the answer they want to submit, before it is marked. The underlying CAS removes any anxiety that alternative correct answers will be missed, which is always a concern when relying on string matching, while good CAS-based systems still leave decisions about what to mark as correct with the question author, for instance deciding whether an unfactorized answer would be considered correct. STACK goes further in also enabling checks for units, precision and the steps used in a calculation (Sangwin and Harjula 2017).

6.5.4 Assessing words, phrases and essays

Many computer-marked assessment systems now include a question type in which students can type their answer as a single word, though the quality of these systems is somewhat variable. The best allow misspelling, if and only if the question author wants this, which allows the assessment of accurate spelling where this is important,

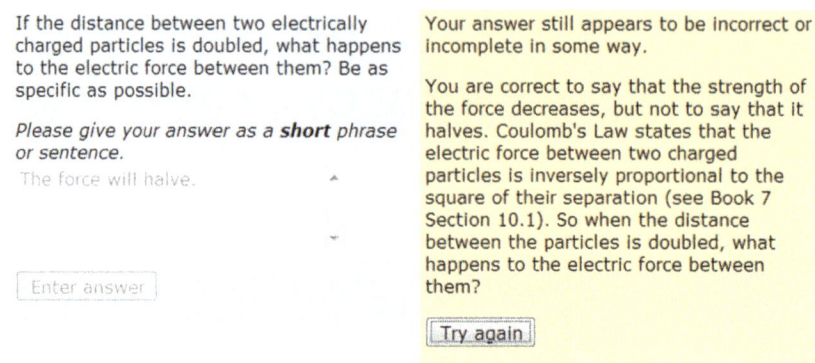

Figure 6.3. An automatically marked question requiring a free-text answer of a few words. This question was written in the PMatch question type, the precursor to Moodle's Pattern Match. Screenshot taken from Moodle™. Copyright Moodle™. Made available under the GPLv3 license.

while also not unfairly penalizing students who misspell common English words, which may be as a result of dyslexia, English not being the student's first language, or a slip which is not relevant to the learning outcome being assessed. Where the correct answer is not a specific technical term, it is also important that the system allows synonyms.

The question shown in figure 6.3 is one that I wrote as part of a project that looked at the automatic marking of free-text answers of phrases and sentences, usually up to 20 words in length. We investigated the use of two contrasting technologies, one making use of artificial intelligence (AI) and one using a 'bag of words' approach, i.e., looking for words (strings of characters) while also considering negation and word order. This meant that, in response to another question, answers such as 'The forces are balanced' and 'There are no unbalanced forces' could be marked as correct, while 'The forces are unbalanced' could be marked as incorrect. Somewhat to our surprise, my colleagues and I found that the relatively simple answer matching was at least as accurate as both human markers and the more sophisticated system, provided that the answer matching was based on human-marked responses from actual students on a similar course (Butcher and Jordan 2010). Despite this finding, subsequent developments in AI and machine learning have suggested interesting avenues for future development (Süzen *et al* 2020).

My colleagues and I have used the same technology that underpins the question shown in figure 6.3, which is available as the Pattern Match question type within Moodle™, to develop a version of the FCI in which some questions are replaced by automatically marked short-answer free-text questions (Parker *et al* 2023). Development work is ongoing, but the tool is approaching sufficient reliability and we hope to use it to gain deeper understanding into conceptual understanding and to investigate some of the known demographic differences in outcome as measured by the conventional FCI.

It is generally considered to be technically easier to obtain accurate automatic marking for essays than it is for short-answer questions. If content is marked at all (which not all essay-marking systems do), simply looking for keywords is often sufficient. Details such as word order and negation are generally found to be less important than is the case for short-answer questions. When essays are marked for style, it is usual to make use of proxies such as sentence and paragraph structure (Shermis and Burstein 2013).

6.5.5 More advanced question types

As technologies develop, so too does the potential for increasingly sophisticated computer-marked assessment questions, for example those assessing mathematical proof. In addition, the growing understanding of the importance of authenticity in assessment has been rewarded by systems such of CodeRunner (Lobb and Harlow 2016), which rather than assessing students' understanding of programming by asking questions about it, asks them to write a simple program, which is evaluated according to whether it works as required.

6.6 Writing questions and feedback

After the assessment has been designed and appropriate question types selected, it still remains to actually write the questions. In this section I offer some tips for question authors, emphasizing the importance of checking your questions and evaluating their performance.

6.6.1 Writing questions

Many of the problems that students experience with computer-marked questions stem from question wording that is in some sense unclear, ambiguous or requires good understanding of a particular culture or language. In addition to avoiding unnecessary contextual information, where possible, I recommend avoiding the use of double negatives. Questions which the author may consider to be 'clever' can all too easily end up assessing an ability to understand the question rather than knowledge or understanding of the course.

Figure 6.4 shows two fictional questions, both deliberately written in 'nonsense language' but illustrating points that I have seen in all too many real questions. Despite the fact that the question has no meaning, it is clear that the correct answer to Question 1 (figure 6.4(a)) is Option B, because of the length of the explanation provided relative to the lack of explanation in the other options. To find the correct answer to Question 2 (figure 6.4(b)) requires an understanding of the English language that should be familiar to those who are native speakers, but maybe not to others; the correct answer is Option C, as this completes the sentence 'The bfeld links to the mnoge by means of a tanag' in a grammatically correct way. All the other options require the final word of the question stem to be 'an' not 'a'; 'The bfeld links to the mnoge by means of a elland' is not grammatically correct.

The examples given in figure 6.4 may seem trivial, but a colleague tells the story of being able to achieve 65% in a multiple-choice assessment despite knowing effectively nothing about the subject.

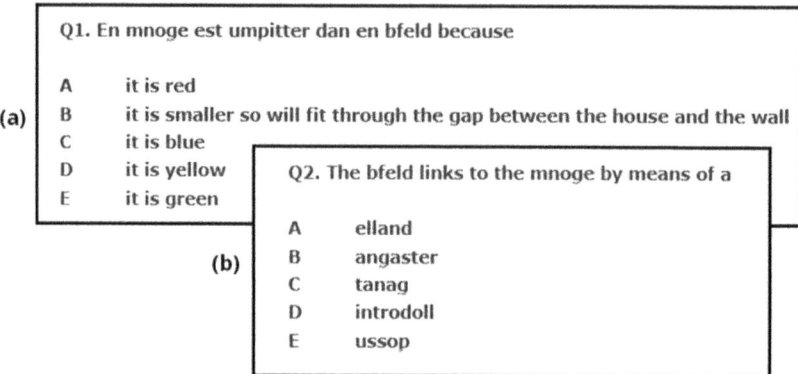

Figure 6.4. Two fictional multiple-choice questions

Checking your own questions should reveal many issues, like these, but it can be particularly difficult for any of us to spot our own mistakes, so I would always advocate checking and rechecking, but also asking a colleague to check your questions.

6.6.2 Distractors, correct and incorrect answers and feedback

When writing multiple-choice questions, the distractors should be plausible answers, preferably based on common misconceptions and mistakes. For constructed-response questions, there is also a need for consideration of both correct and incorrect responses that students are likely to give (or, even better, that students have been observed to give). This enables the question author to ensure that all correct answers are marked as such, and that appropriate feedback can be given.

In much the same way as for the wording of the question itself, it is important that feedback is clear and understood by the student. If they are to learn from it, it is also important that, whenever possible, the feedback makes sense to the student in the context of the answer they have given. Perhaps the largest single source of student frustration is when they are told that an answer is incorrect, but the feedback is too general and does not relate to the student's error. This is particularly irritating to students when their error is minor, or perceived to be. This is exemplified by the following student feedback, received in a survey I conducted:

> I had a go at practice quiz one and when I got to question 2 I got the answer wrong. I spent over an hour going over it and trying to work out where I was going wrong to no avail. Eventually I had to give up, only to discover that my answer was the same as the quiz had except I had expressed my answers to one significant figure more. I was convinced I had lacked understanding of the concept, I was very frustrated and demoralized by this.

6.6.3 Evaluation and iterative design

In addition to asking a colleague to check your questions before use, it is important to monitor them in use, to detect any serious issues and (in summative use) to check that variants of questions are of equivalent difficulty. Various statistical techniques are available to help with this and most computer-marked assessment systems provide basic management information on student performance on different questions and variants. If you are fortunate enough to be able to re-use a question from year to year, you will be able to improve your questions in the light of observed student performance. Figures 6.5 and 6.6 illustrate the effectiveness of one such modification to a question I wrote. The question shown in figure 6.5 did not originally give targeted feedback for the common partially correct answer 'It was formed in a desert'. The answer matching was always acceptable, but simply being told that their answer was incorrect caused much student frustration and, as shown on the left in figure 6.6, very few students were able to correct their answer between and first and second try. The simple addition of the targeted feedback shown in

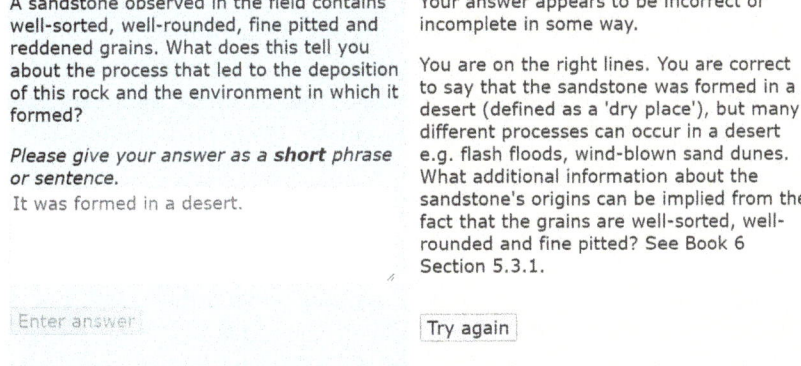

Figure 6.5. A question which has been amended to give targeted feedback on the common partially correct answer shown. Screenshot taken from Moodle™. Copyright Moodle™. Made available under the GPLv3 license.

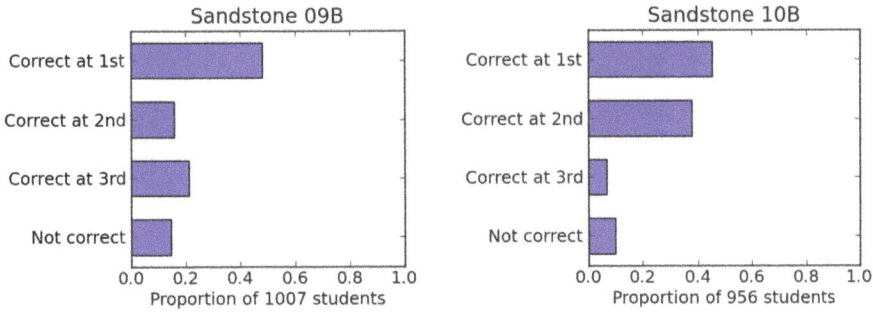

Figure 6.6. The change in question performance as a result of the addition of targeted feedback, between one year (labelled 09B) and the next (10B).

figure 6.5, resulted in a marked improvement in question performance and also to considerably less student frustration.

Analysis of student responses to computer-marked assessment questions can be a rich source of information about student understanding of the topics being taught. Furthermore, analysis of engagement with the system more generally can provide useful information about student engagement with the course as a whole.

6.7 How far should we go?

In this chapter I have outlined the development of computer-marked assessment, pointing towards an increasingly sophisticated future in which we can effectively assess, for example, essays and proof. I have also observed a growing interest in the use of AI to assess students' actual engagement with online teaching activities, rather than considering assessment as a separate event. At one level, this development excites me. However, just because we can do something, it does not mean that we should.

Perelman (2008) famously tricked an essay-marking system by using the proxies that the system was looking for, while demonstrating no understanding of the subject. Particularly in STEM, the subject content is important. Furthermore, essays are intended as a means of communication between two people and therefore I consider that, although automatic systems might support students in developing the relevant skills, a human should be involved in the marking of the final piece of work.

I join the call for variety in assessment (Main 2022, section 6.3.4). Variety supports student diversity and enables appropriate methods to be used for the assessment of different learning outcomes. Computer-marked assessment can motivate and build the confidence of students and provide them with information about their learning. At the same time, it has tremendous potential to free human markers from the drudgery of marking relatively straightforward questions, something that is particularly significant when class sizes are large Teacher time is then freed to help students to interpret the information that the computer has provided, and to deliver effective teaching and the types of assessment that only humans have the skills to do.

References

Ashburn R 1938 An experiment in the essay-type question *J. Exp. Educ* **7** 1–3
Bangert-Drowns R L, Kulik C L C, Kulik J A and Morgan M 1991 The instructional effect of feedback in test-like events *Rev. Educ. Res.* **61** 213–38
Black P and Wiliam D 2009 Developing the theory of formative assessment *Educ. Assess. Eval. Acc.* **21** 5–31
Bloom B S, Engelhart M D, Furst E J, Hill W H and Krathwohl D R 1956 *Taxonomy of Educational Objectives: The Classification of Educational Goals* (New York: McKay)
Bloxham S, den-Outer B, Hudson J and Price M 2016 Let's stop the pretence of consistent marking *Assess. Eval. High. Educ.* **41** 466–81

Boitshwarelo B, Reedy A K and Billany T 2017 Envisioning the use of online tests in assessing twenty-first century learning: a literature review *Res. Pract. Tech. Enhanc. Learn.* **12** 1–16

Bridgeman B 1992 A comparison of quantitative questions in open-ended and multiple-choice formats *J. Educ. Meas.* **29** 253–71

Brown G 2010 The validity of examination essays in higher education *High. Educ. Quart* **64** 276–91

Bull J and Dyson M 2004 *Computer-Aided Assessment* (York: LTSN Generic Centre))

Butcher P and Jordan S 2010 A comparison of human and computer marking of short free-text student responses *Comp. Educ.* **55** 489–99

Cassady J and Grindley B 2005 The effects of online formative and summative assessment on test anxiety and performance *J. Tech. Learn. Assess.* **4** Article 1

Dawkins H, Hedgeland H and Jordan S 2017 Impact of scaffolding and question structure on the gender gap *Phys. Rev. Phys. Educ. Res.* **13** 020117

Dawson P, Sutherland-Smith W and Ricksen M 2020 Can software improve marker accuracy at detecting contract cheating? *Assess. Eval. High. Educ* **45** 473–82

de Oliveira C, Sobral S, Ferreira M and Moreira F 2021 How does learning analytics contribute to prevent students' dropout in higher education *Big Data Cogn. Comp.* **5** Article 64

Dichev and Dicheva 2017 Gamifying education *Int. J. Educ. Tech. High. Educ.* **14** Article 9

Draper S 2009 Catalytic assessment *Br. J. Educ. Tech* **40** 285–93

Gardner-Medwin A R 2019 Certainty-based marking: stimulating thinking and improving objective tests ed C Bryan and K Clegg *Innovative Assessment in HE: A Handbook for Academic Practice* 2nd edn (New York: Routledge) ch 12

Gibbs G and Simpson C 2005 Conditions under which assessment supports students' learning *Learn. Teach. High. Educ.* **1** 3–31

Hestenes D, Wells M and Swackhamer G 1992 Force concept inventory *Phys. Teach.* **30** 141–58

Holmes N 2015 Student perceptions of their learning and engagement in response to the use of a continuous e-assessment in an undergraduate module *Assess. Eval. High. Educ.* **40** 1–14

JISC 2006 *e-Assessment Glossary (Short)* (Bristol: JISC) https://yumpu.com/en/document/read/23670459/e-assessment-glossary-short-version-jisc

Jordan S 2011 Using interactive computer-based assessment to support beginning distance learners of science *Open Learn.* **26** 147–64

Jordan S 2013 E-assessment: past, present and future *New Dir. Teach. Phys. Sci.* **9** 87–106

Kay A E, Hardy J and Galloway R K 2020 Student use of PeerWise: a multi-institutional, multidisciplinary evaluation *Br. J. Educ. Tech.* **51** 23–35

Kitto K and Knight S 2019 Practical ethics for building learning analytics *Br. J. Educ. Tech.* **50** 2855–70

Lobb R and Harlow J 2016 Coderunner: a tool for assessing computer programming skills *ACM Inroads* **7** 47–51

Main P 2022 *Assessment in University Physics Education* (Bristol: IOP Publishing)

Miller T 2009 Formative computer-based assessment in higher education *Assess. Eval. High. Educ.* **34** 181–92

Mitchell T, Aldridge N, Williamson W and Broomhead P 2003 Computer based testing of medical knowledge *7th Int. Comp. Assist. Assess. Conf.*

Montenegro-Rueda M, Luque-de la Rosa A, Sarasola Sánchez-Serrano J L and Fernández-Cerero J 2021 Assessment in higher education during the COVID-19 pandemic: a systematic review *Sustainability* **13** 10509

Nicol D and Macfarlane-Dick D 2006 Formative assessment and self-regulated learning *Stud. High. Educ.* **31** 199–218

Or C and Chapman E 2022 Development and acceptance of online assessment in higher education *J. App. Learn. Teach* **5** 10–26

Parker M, Hedgeland H, Jordan S and Braithwaite N 2023 Establishing a physics concept inventory using computer marked free-response questions *Eur. J. Sci. Math. Educ.* **11** 360–75

Perelman L 2008 Information illiteracy and mass market writing assessments *College Compos. Comm* **60** 128–41

Prisacari A A 2015 The testing effect in general chemistry *MSci Thesis* (Iowa State University, Ames, IA, USA)

Riegel K and Evans T 2021 Student achievement emotions: examining the role of frequent online assessment *Aust. J. Educ. Tech.* **37** 75–87

Ridgway J, McCusker S and Pead D 2004 *Literature Review of E-Assessment* (Bristol: Futurelab)

Roediger H L III and Karpicke J D 2006 The power of testing memory *Persp. Psych. Sci.* **1** 181–210

Sadler I, Reimann N and Sambell K 2023 Feedforward practices : a systematic review of the literature *Assess. Eval. High. Educ.* **48** 305–20

Sands D, Parker M, Hedgeland H, Jordan S and Galloway R 2018 Using concept inventories to measure understanding *High. Educ. Ped.* **3** 173–82

Sangwin C 2013 *Computer Aided Assessment of Mathematics* (Oxford: Oxford University. Press)

Sangwin C and Harjula M 2017 Online assessment of dimensional numerical answers using STACK in science *Eur. J. Phys.* **38** 035701

Shermis M D and Burstein J 2013 *Handbook of Automated Essay Evaluation* (New York: Routledge)

Süzen N, Gorban A N, Levesley J and Mirkes E M 2020 Automatic short answer grading and feedback using text mining methods *Proc. Comp. Sci.* **169** 726–43

Van Gaal F and De Ridder A 2013 The impact of assessment tasks on subsequent examination performance *Active Learn. High. Educ.* **14** 213–25

Winstone N and Carless D 2020 *Designing Effective Feedback Processes in Higher Education* (Abingdon: Routledge)

IOP Publishing

Effective Teaching in Large STEM Classes

Anna K Wood

Chapter 7

Case study 1: an introductory physics course

Ross K Galloway

In this chapter I describe and evaluate Physics 1A, an introductory physics course at the University of Edinburgh. I describe in detail how the course is delivered, including explanations for some of the instructional design decisions. I also give an account of some qualitative and quantitative evaluation of the functioning of the course, and establish that it is effective and performing well. I conclude by offering some personal views on my experience of teaching Physics 1A.

7.1 General context of the course

Physics 1A: 'Foundations' is the first year, first semester introductory physics course at the University of Edinburgh, which is a large, research-intensive university in the United Kingdom. As a Scottish University, Edinburgh follows the Scottish tradition of a broad education, and thus Physics 1A serves two roles: it is the initial course for students following a degree programme in the School of Physics and Astronomy, but it is also available to students on other programmes, who may be required to take it (e.g., for a degree in Geosciences) or who may opt to take it as an elective. Accordingly, the class consists of approximately 75:25 proportions of 'majors' and 'non-majors', though we should note that in order to take the course, all students must have the same basic entry requirements in terms of previous qualifications in physics and mathematics, so 'non-majors' are not less qualified than the 'majors' (though may have differing motivations for studying the course).

The size of the class is typically 300–330 students, of whom approximately 30%–35% identify as female (which is slightly above the sector average). Around 40% of the students are from Scotland, 40% from the rest of the UK, and the remainder from a range of international origins, predominantly in Europe and China. As a highly-selecting university, the programmes are typically over-subscribed by a substantial margin, so most students in Physics 1A possess the highest grades in their previous qualifications in physics and mathematics, so have good prior preparation for the course (as also reflected in their initial FCI scores; see section 7.4.2).

The content addressed by Physics 1A is a recognizably traditional Newtonian mechanics curriculum, featuring kinematics, Newton's laws, forces, energy, linear and angular momentum, and simple harmonic motion. As such, many of the ideas and formulae in the course will be familiar to students from their prior study; however, the course typically re-approaches these ideas in more formal mathematical detail (with a particular emphasis on vectors) and with a much greater focus on problem solving than most of the students are used to. Thus, the course represents an opportunity for students to become enculturated into university-level teaching and learning while on reasonably familiar ground. This relates to a secondary objective of the course, which is to accommodate students coming from a variety of different points of origin and previous experiences, and integrate them into a coherent cohort, while imparting some general study strategies and learning skills that are needed at university.

Physics 1A has a long history of being taught by teachers with a particular interest in pedagogical development and innovation, and has often been in the vanguard of new approaches. It has been taught in the flipped classroom format, largely as described in this chapter, since 2011; thus, it represents an established and fairly stable use of active learning that is well characterized and well understood.

In this chapter I will describe the structure and delivery of the course, discuss how its performance and success have been evaluated, and give some brief personal commentary from the teacher's perspective.

7.2 Structure of the course

Delivery of the course is structured into units of one calendar week as illustrated in figure 7.1.

7.2.1 Personal reading

If we consider a canonical 'week n', in the previous week ('week $n - 1$') the students will have been set clear targets for their personal study. Typically, this will include readings of specified sections of the 'course handbook', a set of hyperlinked online

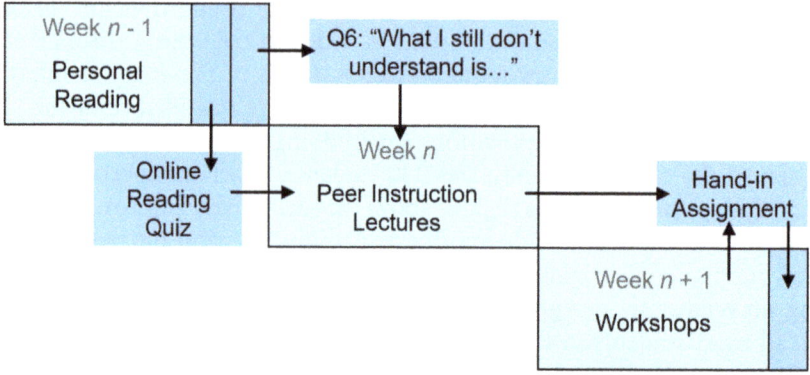

Figure 7.1. Structure of a weekly course segment.

course notes developed specifically for Physics 1A. The course handbook forms a coherent, self-contained set of resources, including direct links to relevant self-test questions, and longer practice problems. The course handbook makes extremely limited use of video content: it is primarily a text-based resource, and not a set of video lectures. While we refer to this week of self-study as 'personal reading', the students are expected to go beyond simply reading: they should work through the mathematical derivations themselves, try self-test questions, and identify any areas that cause them trouble or confusion.

At the end of this week of personal study, the students are set an online quiz on the material they have been reading. This consists of five automatically marked multiple choice questions; a student's score on these questions collectively constitutes up to a 1% contribution to the course's overall assessment. Over the 10 weeks of quizzes, this represents a total contribution of 10%. The choice of this assessment format is primarily driven by pragmatic considerations as the marking requires no human intervention. Since the students are being quizzed on material they have only encountered in self-study, without the direct support of the course teaching team, the level of the quiz questions is intentionally kept to factual statements and simple calculations, rather than deeper conceptual conundrums. The intention is that a lay-person would be unable to answer the questions, but any student who has made a good-faith attempt at the reading should have no major difficulties. We believe the level has been pitched successfully: over all quizzes and over all deliveries of the course, the mean class mark is typically in the mid-80s% range.

However, the weekly reading quizzes are one of the major sources of student disquiet and push-back on the course design: despite the careful writing of the questions to be at a suitable level, and explanations to the students of the distinction between assessment *of* learning, and assessment *for* learning (as these quizzes are intended to be), nevertheless an oft-voiced student sentiment is the complaint that, 'It's not fair to test us on something we haven't been taught yet.' Many students seem to consider that legitimate teaching consists of direct teacher intervention, and do not accept that teacher-provided materials, however carefully structured or scaffolded, constitute a suitable way to learn. One temporary response to this complaint was to remove the 10% course assessment associated with the quizzes: unfortunately, this was not successful as student participation in the quizzes plummeted from high-90s% of the class to only 10%–15%.

At the end of each quiz, there is an additional sixth unassessed free-text question (Q6). This invites the students to tell the course team which topics they found confusing, or on which areas they would like to place a particular focus. Responses to this question (and the previous five MCQs) inform the teacher's plan for the lectures in the following week ('week n'). In practice, however, student difficulties are largely predictable and consistent year-to-year. The content of the lecture sessions are primarily driven by the teacher's experience, pedagogical content knowledge, and overview of the subject. However, the quiz responses do occasionally highlight unexpectedly troublesome topics: the quiz responses drive a perturbation of approximately 10% of the lecture content. This also serves a secondary purpose in communicating to the students that their voices are being heard and responded to, which aids the establishment of dialogue within the course.

7.2.2 Peer instruction lectures

Each week there are three, 50 min whole class teaching sessions. For reasons of convention, we refer to these as 'lectures', however they are not intended to be traditional didactic lecture sessions. They are sessions where the entire class is present (or should be!) in the same room at the same time, typically with a single teacher. The use to which lecture time is put is primarily driven by the decisions of the teacher; this is in contrast to the workshop sessions (see section 7.2.3) where use of time is primarily driven by the decisions of the students. Lectures take place in a raked lecture theatre of traditional architecture.

The guiding principle of the lectures is that they should not feature 'first sight' of any material for the students, which should instead come in the personal reading. This frees the teacher from feeling a responsibility to 'cover' all the material of the course. Instead, the motivation for the lectures is supported exploration of difficult concepts, and modelling expert-like behaviour. The 'skeleton' of the lectures is a series of peer instruction (PI) 'episodes', run in the standard style as developed by Mazur [10]; the PI classroom workflow is described in detail in chapter 2 of this book. These PI episodes are targeted towards known conceptual difficulties, and to allow time for discussion and follow-up, typically each lecture will contain two to four distinct PI episodes.

Figure 7.2, taken from a study which sought to characterize the nature of the activities which take place in flipped classrooms [14], shows the proportion of time spent on different types of activities during Physics 1A lectures. The contribution from PI episodes can be clearly seen in the segments for individual thinking (for the pre-vote), peer discussion, and lecturer feedback. Some component of the segment for lecturer talking will also correspond to the teacher introducing the question. However, this still leaves substantial sections of lecturer talk, and lecturer–student interactions. These interactions correspond to three further major activities that take place during Physics 1A lectures: targeted explanations, worked examples, and experimental demonstrations.

Targeted explanations are re-visitations of topics initially introduced in personal reading that either have been identified by the students as causing trouble, or from the teacher's own experience of difficult topics. Typically this will involve at most 10 min of exposition: students are likely to be more receptive to these explanations in this lecture format than in a didactic lecture, since their attention should have been cued by the pre-reading as to where the difficulties lie. Similarly, worked examples involve the teacher stepping the class through a relevant problem while vocalizing their thought processes and explicitly highlighting expert-like problem solving strategies.

Both of these activities involve 'vicarious interactivity'. This category of activity was identified during the aforementioned characterization study [14]. In short, fully

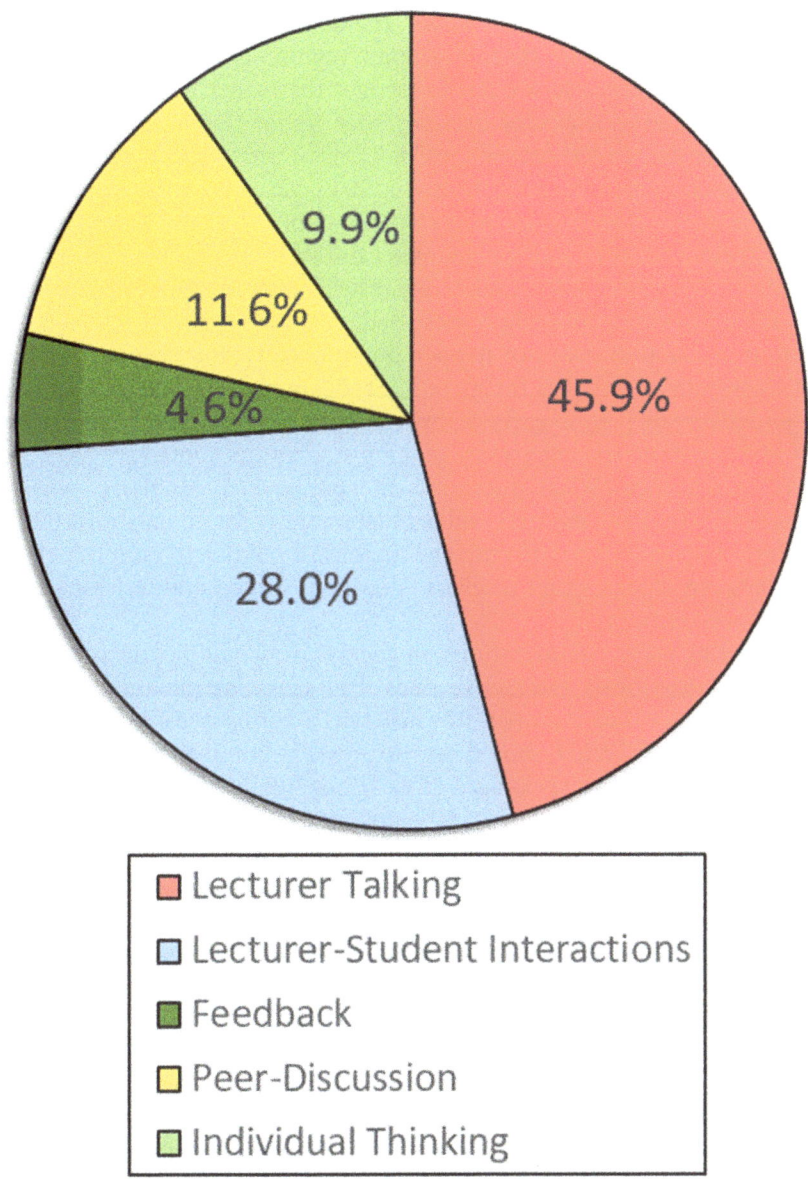

Figure 7.2. Distribution of the use of lecture time across different types of activities, averaged over the semester. Reproduced from [14]. CC BY 3.0.

interactive activities are those (such as PI) where in principle every student present can actively participate. Non-interactive activities are those predominantly led by the teacher, with the student role being more passive. Vicarious interaction sits between these statuses, and might involve prompts from the teacher for student responses, or spontaneous questions from students to the teacher. Thus, while not all students present can or will directly contribute to the discussion, all students can in principle think about how they *would* respond to a question, or what their contribution could be. This middle ground represents a greater level of active engagement than listening to a didactic lecture; the reader might draw an analogy with the different experiences of listening to a seminar, and listening to but not directly participating in the follow-up Q&A. However, there is a natural limit on how many students can be directly involved in vicarious interaction. While the teachers take steps to involve as many voices as possible, for example, ascending the lecture theatre stairs to solicit responses from the back row of seats, nevertheless the customary experience is for a relatively restricted set of bolder or more confident students to be disproportionately represented.

Vicarious interaction does not happen automatically, but requires the establishment of classroom norms that facilitate it. Students pick up 'the rules of the game' quickly and distinctly [13], so early in the semester the course team make a conscious point of inviting, encouraging and welcoming student questions (not 'Are there any questions?' but rather 'What questions do you have?'). Similarly, when asking questions to the class, the teachers will wait (sometimes for an uncomfortable length of time) for a response to be offered. The important point is to establish clearly that dialogue between students and teachers is desired, valued and welcome within the norms of this classroom.

If disregarding other considerations, in general we would favour fully interactive segments over vicariously interactive ones. Full-class interaction can be (and is) incorporated into worked examples (for instance, a voting question on 'What would the next line be?' or 'What should we do next?'), but they are time-consuming compared to the rapid fire exchanges of vicarious interaction, and can disruptively punctuate the flow of an argument. It is also not obvious how to make vicarious discussion resulting from spontaneous student questions into a meaningful full-class interaction. Thus, while full interaction should be the primary skeleton of the session, we do see a place for vicarious elements.

The lecture sessions also make use of experimental demonstrations of physical phenomena. Given the gathering of the class together in a shared physical space, we take the opportunity to demonstrate physical behaviour though tangible experimental means, rather than by recorded video or computer simulation. Live demonstrations are more attention arresting, perhaps because of the element of jeopardy, given that they can go wrong. More importantly, the presentation of the demonstration can be adapted or modified to suit the specific needs of the class, perhaps in response to some vicarious interaction. For example, in a projectile motion demonstration, a student asked 'What would happen if you aimed upwards rather than horizontally?' Such digressions can be explored in a practical demonstration in a way that is not feasible with pre-recorded video. Where the nature of the demonstration allows it, we follow the suggested practice of asking the students to predict the outcome of the experiment beforehand, rather than simple show-and-tell, as the former is generally preferable for conceptual development and recall [1, 11].

7.2.3 Workshops

In the week following a set of PI lectures (i.e. 'week $n + 1$'), the students attend a 2 h workshop session. This takes place in a teaching studio room, of a type similar to that illustrated in figure 7.3. Students work in small groups of six, seated at tables each equipped with a computer and screen to allow easy access to online resources. Some teaching studio spaces are also equipped with paper flipcharts or whiteboards at each table. Members of the teaching team circulate around the room and talk to the groups, assisting them as required.

The focus of the workshop sessions is on supported small-group problem solving. The students are provided with a menu of activities for the session, but the onus is on the students to make productive use of time by focusing on those areas that they find most challenging. The questions go beyond the kind of problem that can be solved with a few minutes of thought, as in the PI questions in the lectures. Here, the students need to engage in extended problem solving, putting the physics concepts into action by solving questions that require detailed working and multiple steps.

The workshops feature a range of activities, which vary week by week but are drawn from these broad categories:

- Course questions: these are the most numerous questions, and represent the central pillar of the workshop activity. They are standard questions requiring students to perform calculations and solve problems on all the topics of the course content. A range of difficulties of question are provided.

Figure 7.3. Illustration of a typical teaching studio room as used for workshop sessions.

- Question of the week: these are open-ended problems situated within a realistic context. They frequently require students to make reasonable estimations of physical quantities or approximations in their calculations. They frequently have no single correct answer.
- Core skills: these are opportunities to practice key mathematical techniques.
- Demo challenges: these feature experimental equipment, and challenge students to predict or explain the outcome of (often counterintuitive) experimental demonstrations.
- Banana skins: these place a spotlight on difficult concepts where experience shows that many students hold incorrect conceptions, i.e., areas where they can 'slip up' easily. Banana skin activities help students to interrogate and deconstruct these alternate conceptions.

Figure 7.4 shows a representative example of a programme of activities from a workshop. Workshops are staffed by an academic who acts as the workshop leader and coordinates the programme of activities, along with a team of typically three postgraduate Teaching Assistants (TA). Approximately 80 students attend each individual workshop session, and the teaching team circulate around the small groups.

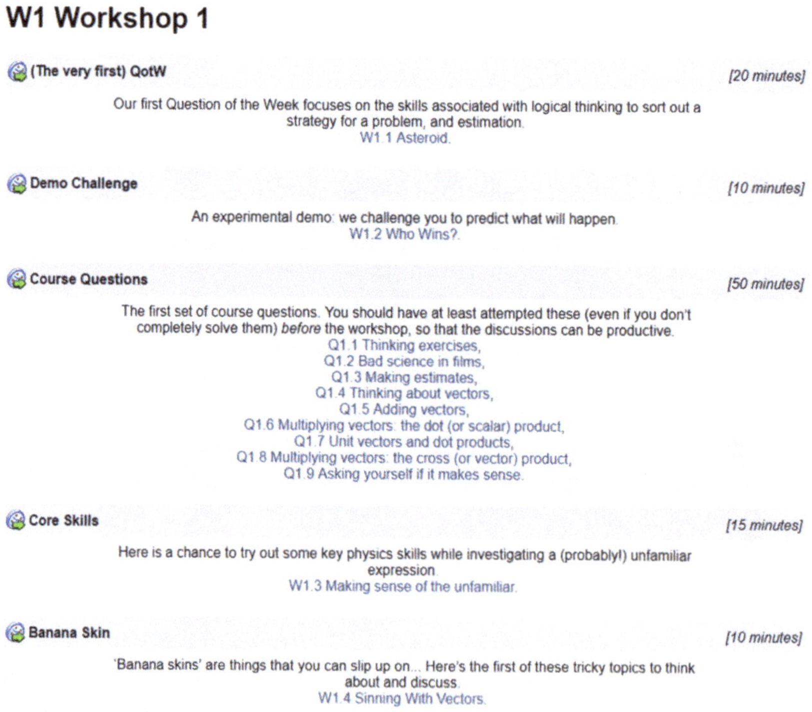

W1 Workshop 1

(The very first) QotW *[20 minutes]*

Our first Question of the Week focuses on the skills associated with logical thinking to sort out a strategy for a problem, and estimation.
W1.1 Asteroid.

Demo Challenge *[10 minutes]*

An experimental demo: we challenge you to predict what will happen.
W1.2 Who Wins?.

Course Questions *[50 minutes]*

The first set of course questions. You should have at least attempted these (even if you don't completely solve them) before the workshop, so that the discussions can be productive.
Q1.1 Thinking exercises,
Q1.2 Bad science in films,
Q1.3 Making estimates,
Q1.4 Thinking about vectors,
Q1.5 Adding vectors,
Q1.6 Multiplying vectors: the dot (or scalar) product,
Q1.7 Unit vectors and dot products,
Q1.8 Multiplying vectors: the cross (or vector) product,
Q1.9 Asking yourself if it makes sense.

Core Skills *[15 minutes]*

Here is a chance to try out some key physics skills while investigating a (probably!) unfamiliar expression.
W1.3 Making sense of the unfamiliar.

Banana Skin *[10 minutes]*

'Banana skins' are things that you can slip up on... Here's the first of these tricky topics to think about and discuss.
W1.4 Sinning With Vectors.

Figure 7.4. Example menu of activities from a workshop session.

Generally, the workshops function well: attendance is good (typically in excess of 90%) and the groups remain on-task and discussing the physics with relatively little prompting from the teaching team. As is not unusual in small-group settings, the dynamics within the student groups are a strong function of their membership, and the TAs need to be alert to one or two very confident students dominating the discussion, or quieter students being sidelined. A judicious interjection or two from a TA can help with this, but it remains a challenge. In general, the TAs are kept well occupied by questions and discussion with the groups, with little unengaged time.

7.2.4 Overall structure of the course

Figure 7.5 illustrates how the three types of teaching activity slot together to form a semester of instruction. Distinct weekly units of content are shown in consistent colours. In a canonical week n, students are conducting personal reading on topic C (yellow), attending PI lectures on topic B that they prepared the previous week (blue), and attending workshops on topic A that they first encountered the week before that (green). Thus, while primarily motivated by pragmatic scheduling considerations, the structure of the course delivery incorporates elements of spaced and interleaved practice [12]. Overall, the course consists of ten weekly units of content plus an introductory orientation week to give an 11 week semester.

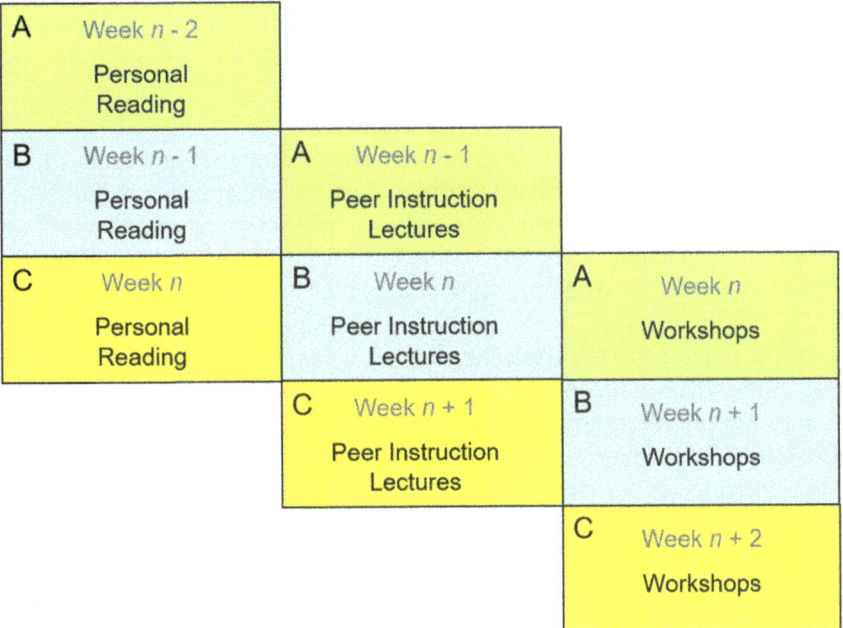

Figure 7.5. Overall structure of the course, showing overlapping components of any given 'week n'.

7.3 Assessment

As previously discussed, 10% of the course assessment comes from the weekly online quizzes. A further 10% comes from other continuous assessment tasks. These are:

- Two hand-in assignments. These are sets of physics problems that students attempt by themselves, in their own time, and hand-in for marking. There is a period of five days between the release of the questions and the submission deadline. The first assignment consists of five 5-mark short questions, similar to the short questions that will appear in the main exam, and tests a range of topics in a fairly routine way. The second assignment consists of one 25-mark long question, again similar to a long question in the main exam, that develops a more limited set of topics to greater depth and with a consistent theme running through the question. Each hand-in assignment contributes 3% towards the course assessment.
- A class test. This is scheduled after the second assignment, and also features a long 25-mark exam-style question. This is done under realistic exam-style conditions, time-limited and in an invigilated room. Thus, it gives students an opportunity to experience an exam-style environment but in a more low-stakes situation, as the class test only contributes a further 3% towards the course assessment. Thus, the combination of the two hand-in assignments and the class test sequentially raises the expectations placed on the students, allowing them to build steadily towards the challenge of the main exam.
- The final 1% of the continuous assessment comes from an online task [9] using the PeerWise multiple choice question authoring and sharing system [4]. This task motivates the students to contribute to and use a peer-generated bank of practice questions that remains available as a study resource throughout the semester.

The remaining 80% of the course assessment comes from the final exam, which has a duration of 2 h and is conducted in a traditional-style invigilated exam hall. The exam consists of 25 marks drawn from short-style 5-mark questions and 50 marks drawn from long-style questions (two of 20 marks and one of 10 marks). The exam is structured as an 'open notes' exam: students can take into the exam hall anything they like that is written on paper, *except* published, bound textbooks. The justification for this is that it is the process of selecting and developing their own resources that is valuable, rather than of having them accessible during the exam *per se*. Developing a set of exam resources requires the students to interrogate the structure of the course and decide which areas are most important, alongside identifying topics with which they have difficulty. In the case of a textbook, the book authors have already carried out that process, robbing the students of the value of doing it themselves.

7.4 Evaluation

There are a number of ways in which the success (or otherwise) of the course can be evaluated. For quantitative approaches, it is often useful to compare the mean class score on some assessment instrument before instruction, ⟨Pre⟩, with that after instruction, ⟨Post⟩. A useful metric is the so-called mean normalized gain, ⟨g⟩. This is defined by

$$\langle g \rangle = \frac{\langle \text{Post} \rangle - \langle \text{Pre} \rangle}{100\% - \langle \text{Pre} \rangle}.$$

Essentially, the mean normalized gain gives the mean improvement of the class (i.e. ⟨Post⟩ − ⟨Pre⟩) expressed as a fraction of the maximum possible level of improvement (i.e. the 'amount of headroom', 100% − ⟨Pre⟩). This metric is also sometimes referred to as the Hake gain after its originator [6].

Mathematically, ⟨g⟩ is bounded above by 1.0, which represents the maximum possible normalized gain, where all students give the correct answer in the post-test. A ⟨g⟩ of 0.0 corresponds to no improvement at all, and deterioration results in a negative ⟨g⟩ value. Hake identified three categories of normalized gain: low gain (⟨g⟩ < 0.3), medium gain (0.3 ⩽ ⟨g⟩ < 0.7), and high gain (0.7 ⩽ ⟨g⟩ ⩽ 1.0). These categorizations are essentially arbitrary, but have received widespread traction in the literature.

7.4.1 Peer instruction episodes

Figure 7.6 shows the normalized gains that occur between pre-discussion votes and post-discussion votes for all PI episodes in a representative delivery of the course; not shown are those instances where a sufficiently large fraction of the class chose the correct answer in the pre-vote, so that the discussion phase was skipped. As can be seen, there is a wide range of ⟨g⟩ across different questions. Those with a high ⟨g⟩ are functioning well; for those with poor ⟨g⟩, the course teaching team may simply discard the question for future iterations of the course, or more probably refine it

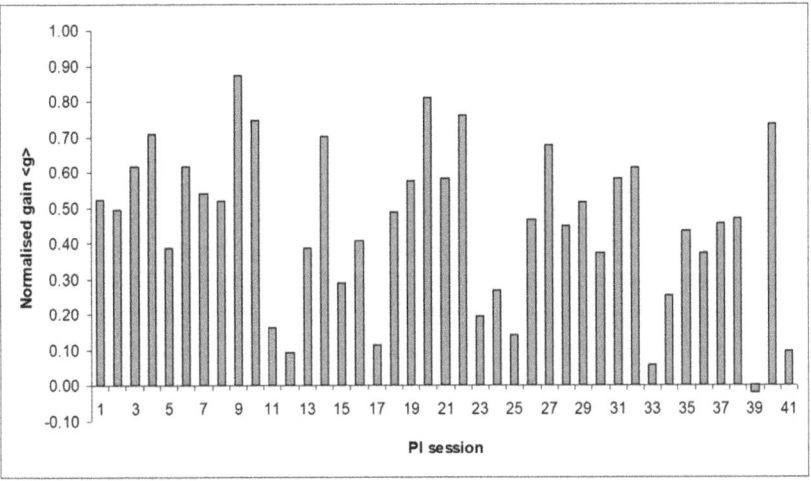

Figure 7.6. Distribution of normalized gains between the pre-discussion and post-discussion votes for a series of PI episodes from a representative delivery of the course.

to try to reduce any ambiguities, misleading aspects, or unhelpful distractors. These refinements are informed by the discussions with the students during the lectures. Over all deliveries of the course, the mean $\langle g \rangle$ for the PI questions is typically around 0.4.

PI session 39 in figure 7.6 shows a small but negative normalized gain: this is an instance where the proportion of students in the class choosing the correct answer option has decreased following the PI discussion, i.e., discussing the question with other students has 'made things worse'. In such cases, the teacher must step in to clearly explain the correct reasoning, sometimes conducting a follow-up vote to make sure that their explanation has been taken on board (at least in the moment) by the students. Such instances of negative gain are, in our experience, very rare: figure 7.6 is representative in that negative gains appear typically once or twice in the entire course. Thus we see that PI is generally highly successful in the sense that, in most instances, the proportion of students selecting the correct answer increases appreciably following discussion with their peers.

7.4.2 Force concept inventory

The force concept inventory (FCI) [8] is a standard diagnostic test instrument in basic mechanics. It consists of 30 multiple choice questions on concepts related to force and motion. Because the syllabus of introductory courses in Newtonian mechanics tends to be highly consistent across different universities, it is possible to deploy the FCI identically in multiple institutions, allowing meaningful comparison between courses in universities across the world. Chapter 6 of this book gives more discussion about the employment and utility of concept inventories in the most general sense.

We deploy the FCI in Physics 1A in the standard way, that is in a pre/post comparison. The students answer the FCI in week 1 of the semester, before any instruction has taken place (and the correct answers are not revealed to them), and then again at the end of the course in week 11. This allows us to calculate a mean normalized gain on the FCI for the class, and thus evaluate the performance of the course. This is what Hake did in his seminal 6000-student investigation of interactive engagement versus traditional methods [6].

Table 7.1 summarizes the FCI data from all presentations of the course in its current format. With the exception of two recent years, which were directly affected by the COVID-19 pandemic, the normalized gain lies clearly within Hake's range for medium gain: this is consistent with Hake's findings for other introductory physics courses using an active learning approach, and higher than those using a traditional didactic approach, which typically have $\langle g \rangle$ in the low range [6].

Also notable are the high pre-test scores in table 7.1: whereas Hake found a mean pre-score of 48% for university-level courses, over all deliveries of Physics 1A we see a mean pre-score of 21.26, or 71%. This means that our students arrive comparatively better-prepared in Newtonian mechanics, leaving relatively little 'headroom' in the FCI. While normalizing the gain addresses this issue somewhat, discretization effects are also relevant: an average student has only eight initially-incorrect

Table 7.1. Force concept inventory data for all deliveries of Physics 1A in its current format: pre-instruction mean score out of 30; post-instruction mean score out of 30; mean normalized gain.

Academic year	Pre-score (/30)	Post-score (/30)	$\langle g \rangle$
2011–12	19.22	25.14	0.55
2012–13	*	*	0.51
2013–14	20.75	24.94	0.45
2014–15	*	*	*
2015–16	18.47	24.87	0.56
2016–17	23.84	26.54	0.44
2017–18	22.09	25.30	0.41
2018–19	22.13	25.19	0.39
2019–20	22.13	25.48	0.43
2020–21[†]	21.70	24.08	0.29
2021–22[†]	21.10	23.80	0.30
2022–23	21.19	24.53	0.38

[*] Data unavailable.
[†] Online teaching during the COVID-19 pandemic.

questions on which to improve, such that a few persistently challenging questions can have a disproportionate numerical effect on the realized $\langle g \rangle$. Arguably, the FCI is approaching the ceiling of its utility in Physics 1A, but we have maintained its use because of the ubiquity of the FCI as a comparator across the discipline.

The FCI measures only one limited aspect of Physics 1A, that of conceptual understanding of forces and motion, but by this measure the course is performing comparably to other, similar courses that also employ an active learning approach, and better than is typically found in more traditionally delivered courses.

7.4.3 Student perspectives

Figure 7.7 shows the relative distribution of student responses to an anonymous survey question following the initial delivery of Physics 1A in flipped format. The distribution is strongly skewed towards favouring the Physics 1A approach. These students would have been taking at least one mathematical methods course, and probably one other course, in a traditional format that they could use as a point of comparison. Similar follow-up surveys in subsequent years repeat this broad pattern.

In addition to quantitative surveys such as these, the course is routinely evaluated by students using free-text responses. These tend to mention the level of interactivity, including the PI episodes, as a positive feature, along with the physical demonstrations during lectures. Positive comments also frequently mention the quality of the explanations given in the course. As mentioned in section 7.2.2, we attribute this positivity to the fact that the explanations are targeted, and the students are cued to be receptive to them.

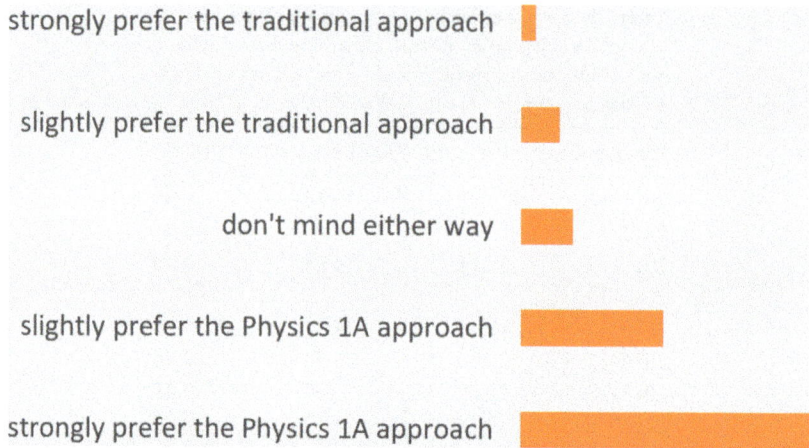

Figure 7.7. Distribution of student preferences from the first (2011–12) delivery of the course in its current format.

Also evident from figure 7.7 is that some non-negligible fraction of students prefer a more traditional approach. The most frequent nature of free-text comment from these students is some variant of 'You didn't actually teach us anything,' or 'We had to do everything ourselves'. This is despite the data in figure 7.2 showing that in actuality, nearly half of lecture time is taken up by the teacher speaking directly to the class. These perceptions persist year-to-year, and probably represent a group of students who do not accept the validity of input by peers to their learning, or who value an authoritative voice in the classroom.

While the majority of students are enthusiastically receptive, resistance to active learning approaches is not unusual [5]. We make explicit attempts to justify and explain our instructional design philosophy during the introductory orientation week: as can be seen from the structure of the course described by figure 7.5, in the first week of the semester the students are carrying out personal reading on the initial topics of the course. Thus, the lecture sessions in the first week cannot rely on this material. Instead, we use the time to explain the course design in detail, introduce a research-informed physics problem solving strategy [3, 7], and show the students published studies and data, e.g., [2], that support this pedagogical strategy. However, despite this, some students continue to be dissatisfied with it.

7.5 Teacher perspectives

In this section I will give a personal reflection on the experience of teaching Physics 1A. I have never viewed the flipped classroom as an end in itself; rather, it is the means by which classroom time can be made available to allow the use of active learning techniques. Freedom from the tyranny of 'coverage'—that is, not feeling compelled to have to say everything of importance out loud in the classroom—transforms my experience of teaching sessions. Space and time become available to meaningfully respond to unexpected occurrences: if a PI question reveals that the class is having more trouble with some particular concept than might be expected, there is scope to spend longer exploring it, with additional follow-up questions, further examples, or open discussion. If a student spontaneously asks a relevant and interesting question, it can be addressed in detail. Worked examples can be modified in real time based on vicarious interactions with the students, without worrying about 'not getting through everything'.

Such 'in the moment' teaching is interesting, exhilarating, and always fresh and rewarding. However, one consequence is that it is impossible to fully plan and completely prepare for any given teaching session, as one never completely knows in which direction it might go. The teacher needs to be sufficiently assured that they are on top of their subject to confidently respond to unexpected interventions, and be willing to go into the classroom not knowing what will happen: this can be off-putting for some teachers, particularly those who are unused to this style of class. It may not be a good fit to the teaching preferences of all teachers.

However, I have found this style of teaching to be a 'one way gate': having tried it, I would not be able to see myself returning to a more didactic approach.

7.6 Summary

Physics 1A: Foundations has been taught in a flipped classroom, active learning format for over a decade. As such, it represents an example of a stable, mature, well-understood and well-characterized course of this type. The course consists of a tightly-structured combination of: directed personal study; large-group, PI driven lectures; small-group, problem solving workshops; and integrated assessments. The course is generally popular with students, and by a number of quantitative measures is performing well. Its effectiveness in developing conceptual understanding is comparable with similar active learning courses in other institutions. It is also rewarding and enjoyable to teach.

Acknowledgements

Physics 1A has relied on many dedicated teams, too numerous to name in full, but including postgraduate Teaching Assistants, course administrators, technicians, learning designers, and software developers, among others. However, I would like to particularly acknowledge Simon Bates, Richard Massey, Will Hossack, John Loveday and Alexander Morozov.

References

[1] Crouch C H, Fagen A P, Callan J P and Mazur E 2004 Classroom demonstrations: learning tools or entertainment? *Am. J. Phys.* **72** 835–8
[2] Crouch C H and Mazur E 2001 Peer instruction: ten years of experience and results *Am. J. Phys.* **69** 970–7
[3] DeMuth D 2022 *A logical problem solving strategy* https://groups.physics.umn.edu/physed/Research/CRP/psintro.html (accessed 23 December 2022)

[4] Denny P, Luxton-Reilly A and Hamer J 2008 The PeerWise system of student contributed assessment questions *Proc. 10th Conf. Australasian Computing Education* 78(ACE'08) (Sydney: Australian Computer Society, Inc.) pp 69–74

[5] Deslauriers L, McCarty L S, Miller K, Callaghan K and Kestin G 2019 Measuring actual learning versus feeling of learning in response to being actively engaged in the classroom *Proc. Natl. Acad. Sci.* **116** 19251–7

[6] Hake R R 1998 Interactive-engagement versus traditional methods: a six-thousand-student survey of mechanics test data for introductory physics courses *Am. J. Phys.* **66** 64–74

[7] Heller K and Heller P 1995 The Competent Problem Solver: A Strategy for Solving Problems in Physics *Calculus Version* 2nd edn (Minneapolis, MN: McGraw-Hill)

[8] Hestenes D, Wells M and Swackhamer G 1992 Force Concept Inventory *Phys. Teach.* **30** 141–58

[9] Kay A E, Hardy J and Galloway R K 2020 Student use of PeerWise: a multi-institutional, multidisciplinary evaluation *Br. J. Educ. Technol.* **51** 23–35

[10] Mazur E 1997 *Peer Instruction: A User's Manual (Series in Educational Innovation)* (Englewood Cliffs, NJ: Prentice-Hall)

[11] Miller K A, Lasry N, Chu K and Mazur E 2013 Role of physics lecture demonstrations in conceptual learning *Phys. Rev. ST Phys. Educ. Res.* **9** 020113

[12] Rohrer D and Taylor K 2007 The shuffling of mathematics problems improves learning *Instr. Sci.* **35** 481–98

[13] Turpen C and Finkelstein N D 2010 The construction of different classroom norms during peer instruction: students perceive differences *Phys. Rev. ST Phys. Educ. Res.* **6** 020123

[14] Wood A K, Galloway R K, Donnelly R and Hardy J 2016 Characterizing interactive engagement activities in a flipped introductory physics class *Phys. Rev. Phys. Educ. Res.* **12** 010140

IOP Publishing

Effective Teaching in Large STEM Classes

Anna K Wood

Chapter 8

Case study 2: tailored active blended learning in a foundation year chemistry module

Simon J Lancaster, Daniel Elford and Eleanor Gill

In this chapter we discuss the *Introductory Chemistry* module at the University of East Anglia, which caters to a cohort that is diverse in both experience and career aspiration. Our approach is a tailored active blended learning programme, combining online tests to gauge the level of support needed by each student, 'gamification'—gentle competition amongst student sub-cohorts, and a standard peer instruction model. This is further complemented by the crowd-sourcing of potential distractors to surface otherwise unanticipated misconceptions. Peer interactions are extended beyond the immediate neighbours using Padlet as a synchronous networking and discussion tool. The enhanced conceptual insight is further reinforced through more demanding follow-up assessments.

8.1 Introduction

8.1.1 The context

Introductory general chemistry is a pre-requisite of many science and health degree programmes. Chemistry educators therefore often face the challenge of delivering a grounding in the discipline to students who would not have chosen to study the subject. The prior experience of these students has often been of chemistry taught as a litany of facts and not as a fundamentally conceptual discipline. This presentation as incoherent information to be rote-memorized can negatively shape perception.

    This case study concerns teaching a module entitled *Introductory Chemistry* to a Foundation Year (FY) Cohort in the Faculty of Science at the University of East Anglia (UEA). UEA is a dual (research and teaching) intensive institution based in the provincial city of Norwich in the East of England. The FY is a FHEQ Level 3 course equivalent to A Levels in England, and the first year of a Scottish, North American or Australian degree.

In England, the FY presents an alternative to the traditional A Level entry route for science undergraduate students. The FY attracts students with a variety of prior educational backgrounds. Some of our students are changing academic direction, having previously studied other subjects at A Level. Many have hitherto had limited educational opportunities. While for others, the FY represents another chance at a chosen university programme denied to them by their existing A level grades.

Thus, students who wish to study subjects as diverse as ecology and pharmacy, but would never have voluntarily chosen a chemistry course, find themselves compulsorily enrolled on an introductory chemistry course, where they are alongside students wishing to go on to biochemistry, who have previously studied chemistry to A Level. This presents significant challenges for the educators. We have sought to address these through a tailored blended learning provision, placing peer instruction (Mazur 1999), in which the different cohort identities engage in low stakes competition, at the centre of our pedagogical approach. Cumulatively, while chemistry with a foundation year is sparingly popular, adding in all FY science degrees, for which chemistry is a facilitating subject, sums to a large enrolment, for UEA science, at approximately 200 students.

8.1.2 Peer instruction in chemistry

The case for active learning is made in chapter 2 and applies equally well to chemistry. Active learning is time intensive and challenges the educator to make space for it in their packed curriculum. One such approach to making that space is flipped teaching, which seeks to move some of the transmission of information to asynchronous delivery. The term 'flipped teaching' is attributed to the High School chemistry teachers, Bergmann and Sams (2012). The extensive chemistry education literature on flipped teaching and the associated active learning approaches has been reviewed by Seery (2015). Given the common dependence on conceptual understanding, and its championing by the active learning community in physics, we are surprised there are so few reports of peer instruction being employed in chemistry higher education (Schell and Butler 2018).

Our adoption of peer instruction was inspired by the example of the first year physics at the University of Edinburgh, which is described in chapter 7. In essence, we have adopted a similar pedagogical approach and have reported it in a previous book chapter (Lancaster 2019). The chapter herein focuses on implementation and modification of the peer instruction pedagogy, made to address concerns about engagement and add value for our very diverse students: the tailoring of the flipped learning provision; crowd-sourcing potential answers and the gamification of the polling experience.

8.2 Tailored active blended learning

8.2.1 The rationale

The central premise of our approach is that the student experience should be modified to reflect their prior experience (Chamberlain *et al* 2021). As discussed above, the student intake to the Introductory Chemistry module is very mixed. Some know next to no chemistry and others completed an A level course just a few months earlier. A simple flipped learning approach would require all students to asynchronously study preparative materials before attending a peer instruction teaching session. Our module begins with a very basic introduction to states of matter. If we require our students to review already familiar materials at the very beginning of the module, we will rapidly exhaust the initial student enthusiasm and even their confidence in our academic judgement and with it our authority to direct student activity. Our experience is that this has negative consequences, as we proceed to progressively more complex concepts during the semester. We need to tailor the extent of study to the needs of the individual students. Since the preparative work takes place asynchronously online, we describe this strategy as tailored active blended learning. The lack of correlation between what students know and their self-confidence is well documented (Dunning *et al* 2003) and so they require some direction in which optional resources to study.

8.2.2 A weekly structure

Introductory Chemistry is divided into weeks of study, each of which has a similar pedagogical structure as depicted in figure 8.1.

8.2.3 Introductory quiz

Every week begins with an introductory quiz, presented on the virtual learning environment, the purpose of which is simply to determine the familiarity of the student with the week's subject matter. The notion that the success of active learning hinges on the quality of the question posed is a recurring theme of this text.

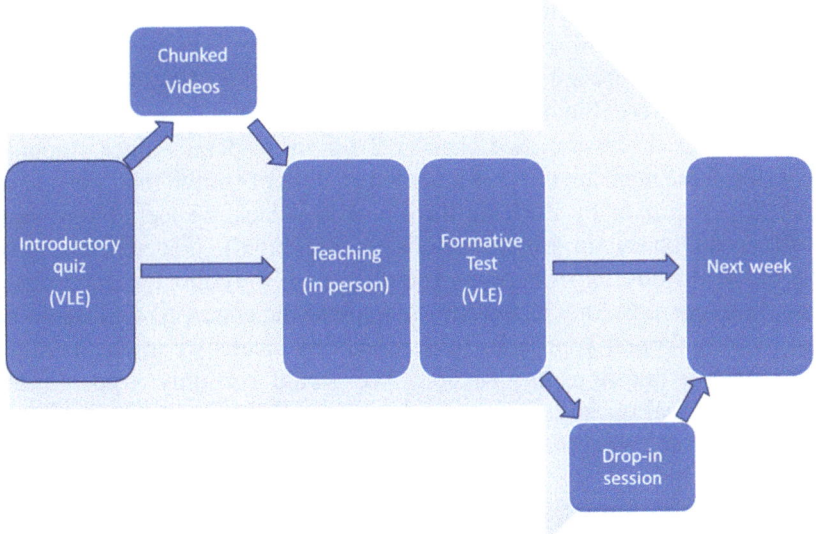

Figure 8.1. The weekly module structure.

However, the introductory quiz is not seeking to facilitate learning, merely to assess a student's prior knowledge and therefore the questions can be simply recall in nature. Indeed, at this point in the weekly sequence there is a need to ensure that the student is not intimidated by the conceptual demands of the week's study.

The results of the introductory quiz direct how much preparation the student needs to do. The automated feedback tool on the Blackboard VLE recommends study materials when the student gets a question wrong. In our programme the introductory quiz is purely formative, and we do see a reduction in the proportion of students taking it during the semester.

8.2.4 Chunked videos

In our case, the choice of recommended study materials has been swayed by the availability of chunked video resources. We define a *chunked video* as a short (less than 6 min) video segment focusing on a specific theme or concept. They began as recordings of teaching sessions from previous years, which have been edited and pared down to a core message. Chunked videos could be replaced by notes, podcasts, or textbook passages. While students are free to view all the available videos, we encourage them to focus on those directed by the introductory quiz or where they know they lack confidence.

8.2.5 Synchronous teaching

The next step in our weekly sequence is in-person teaching. Given the large class size, this typically takes place in traditional lecture theatre spaces. Formally, attendance is compulsory, and no student is excused attendance, regardless of their score in the introductory quiz. However, we do not award academic credit for attendance. Our objective is to foster a conceptual approach to chemistry, and we believe all the students on the module will benefit from attendance.

Our pedagogical approach is modelled upon the work of Eric Mazur, which uses peer instruction—a combination of questions designed to test conceptual understanding and peer discussion. See chapter 2 for more detail. Since students have demonstrated some measure of prior knowledge, either through their performance in the introductory quiz or by studying the chunked videos, we feel able to accelerate the transmission phase (in which material is presented). Nevertheless, given the diversity of the attitude, experience, and confidence of the cohort we have chosen not to forego the exposition (in which we try to explain the concepts) and we do not rely upon an entirely flipped approach. In practice this means we might have time for only four or five questions per session. We would certainly argue that a few, genuinely conceptual questions, that spark constructive discussion, can have much greater influence on conceptual understanding than a greater number of less incisive queries (figure 8.2).

The peer instruction process and the importance of the question are discussed in chapters 2 and 4 respectively. Throughout the development process, the questions were reviewed and checked for content validity to ensure student attention was

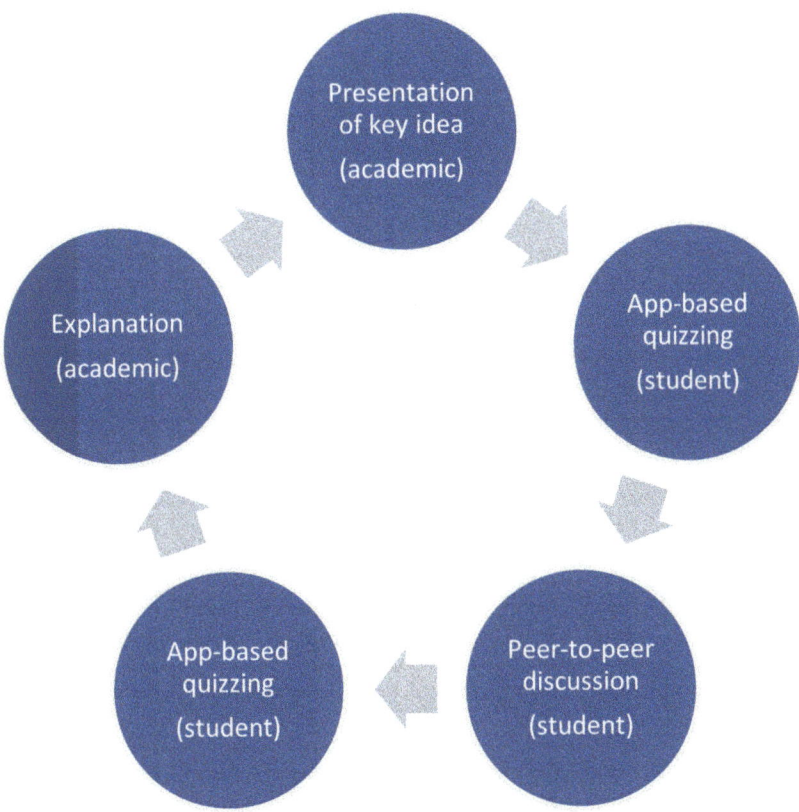

Figure 8.2. Our implementation of the peer instruction process.

focused on critical concepts key to addressing specific learning goals. To satisfy these requirements, we used the following six criteria for each question (Newbury 2013):

(i) **Clarity.** Students should waste no cognitive resources understanding the requirements of the question.
(ii) **Context.** The question should be appropriate for the learning material.
(iii) **Learning outcome.** The question should allow students to demonstrate that they grasp the concept.
(iv) **Distractors.** Distractors should be plausible solutions to the question.
(v) **Difficulty.** The question should not be too easy or too hard.
(vi) **Stimulates thoughtful discussion.** The question should engage students and incentivise thoughtful discussion.

We stress the anonymity of the individual student answers. Where our practice differs from classical peer instruction is in our sharing of the results of the first round of polling with the students. In our judgement, the value of students being able to observe the peer instruction process moving the cohort collectively towards the correct answer builds student confidence in the pedagogy and their commitment to the process. Questions in which one answer is heavily favoured in the first round, risking prejudicing subsequent behaviour, are already failing against criteria (v) and (vi).

Recognizing the pivotal importance of the peer-to-peer discussion phase, much of our effort has been focused on encouraging student discourse. Considerable academic capital is invested in explaining, cajoling, corralling and near pleading with students to engage both in polling and discussion. Persistence prevails and students do become used to the notion of lively debate in class. So much so, that we find it necessary to remind the students that the first round of polling should be their individual opinions.

Our experience confirms the importance of the final step in the peer instruction process: the teacher's explanation of the correct answer. It is possible for students to arrive at the right answer for the wrong (and very convincing) reason. Furthermore, when students revise the material, they will not have access to their peer discussions. This part of the teaching materials, class notes or recordings are therefore highly valued by students.

8.2.6 Formative test

To exercise the new-found conceptual understanding and confidence in the week's topic and to ensure they are ready for examination-level assessment, students are directed to take an automated test. The questions in the end-of-week test are as academically challenging as any we pose on the module and assess understanding and application. This contrasts with the introductory quiz questions, which simply probed familiarity. Given the formative status, we have invested a lot of effort and energy in encouraging students to engage, focusing on the direct relevance to the final examination. Despite this, the proportion taking the formative test does decrease during the semester.

8.2.7 Workshops

Students who are not content with their results in the formative test are presented with a range of follow-up options. They can potentially review either the chunked videos or the recording of the week's lecture. We provide a drop-in workshop, on the Friday afternoon. These workshops are promoted as opportunities for students to bring any questions they might have on the week's teaching. Pointedly, they are told that we will proceed not at a pace dictated by the material, but one dictated by the least confident student. The challenge is to manage attendance, such that the *worried well* do not feel obliged to attend and that students who need additional help are not intimidated. From an academic perspective, we aim to have a much less diverse cohort in the workshops. The attendance is typically less than 10% of the total enrolment.

Sections 8.2.3–8.2.7 outline a typical week on the module and the series then repeats with fresh topics the following week. Weeks 6 and 12 are dedicated to a formative course test, and review and revision sessions respectively. The focus is on examination technique and preparation.

8.3 Class-sourcing misconceptions

Misconceptions and misunderstandings are a scourge of teaching in the physical sciences and much of our challenge is in identifying and correcting them. In flipped teaching, it is common practice to include a supplementary question in any pre-lecture activities, enquiring what students would like to focus upon, in order that the teacher can tailor the synchronous session accordingly. In our experience, students are not particularly forthcoming and students who do volunteer a topic, not particularly representative of their cohort. In part this may be apathy, but it doubtless also reflects the intrinsic nature of misconceptions—students do not know they hold them. Typically, we rely on our experience and those of colleagues to select distractors that target misconceptions and allow us to confront them. However, there will be instances where students are harbouring misconceptions we have not anticipated. How can we surface these?

Figure 8.3 describes a variation on the classic peer instruction process. Instead of initially presenting the peer instruction question as a multiple-choice problem, it is presented as a free text question. Students anonymously respond with their

Figure 8.3. Modified peer instruction process, including class-sourced misconceptions.

proposals and in so doing, they (hopefully) reveal the correct answer and the most prevalent misconceptions. We then construct, on-the-fly, a classic peer instruction multiple-choice question and enter the normal peer instruction cycle.

The benefits of this approach are best illustrated by an example. *'Where does most of the mass of a mature oak tree come from?'* is an often-used peer instruction question. Normally the audience are presented with the four options: (a) the soil; (b) the acorn; (c) the rain; and (d) the air. When we posed this question as free text, a significant number of respondents wrote 'the Sun'. The misconception that the Sun, which provides the energy for photosynthesis, was providing the mass for the oak tree was not being addressed in the original question. In the spirit of peer instruction, the correct answer is (d)—carbon dioxide from the air is the major source of cellulose and carbohydrates in the (dry) mass of the tree. Where the definition of dry launches further impassioned conceptual discussion amongst experts in the field.

8.4 Fostering and utilizing sub-cohort identities through gamification

8.4.1 Extending in-person peer discussion with online tools

The COVID-19 pandemic presented a significant challenge to our active-learning focused pedagogy. The first semester of the 20/21 academic year saw all large group teaching at UEA presented online using the Blackboard Collaborate tool. Our intention was to replace the near neighbours of the lecture theatre with members of breakout groups. In practice this worked very poorly. The breakout groups were created at random and were populated by students who did not know each other and were reluctant to share web cameras or even turn on microphones. In theory the educator could navigate breakout rooms but in practice this took much too long compared to the normal discussion phase of peer instruction. So we began to look for alternative approaches.

In parallel to the development of the pedagogy outlined in section 8.2, we have been attempting to improve engagement with our virtual learning environment, and foster a culture of asynchronous interactions between students, similar to the social learning on Massive Open Online Course platforms (Anderson *et al* 2015). It has proven extremely difficult to convince students to enter asynchronous discussions no matter how interesting, relevant or representative we and our undergraduate collaborators tried to make them. We had begun to despair of ever successfully utilizing the discussion board feature.

We achieved a minor breakthrough by using the discussion boards to synchronously host the peer instruction discussions. A discussion board thread was prepared in advance for each of the peer instruction questions. Instead of sending online students to breakout rooms, they were directed to the Blackboard discussion thread. Where they were encouraged to present their arguments in favour of a particular answer option and debate the merits with their peers in real time as part of the online session. It proved much easier to monitor and encourage engagement with the discussion boards than the breakout rooms. Furthermore, engagement of the synchronous student body with the discussion boards was much greater than it had ever been asynchronously. However, the number of students actively taking part in the discussion was still much less than for in-person peer instruction. We were grateful to return to campus for large group teaching in the 21/22 academic year.

Most students on Introductory Chemistry also take Introductory Biology. From 20/21 onwards that module has employed a different software tool, Padlet, as the platform for asynchronous discussion and some in-class evaluation. Padlet is much more aesthetically pleasing than Blackboard discussion boards, and given the added benefit of student familiarity, we now favour it over Blackboard.

8.4.2 Extrinsic motivation through badging and gamification

The Blackboard platform can allocate badges to students for completing activities or reaching the threshold in tests. In practice we found this approach rather ineffectual in driving engagement. While it proved a simple way for the module organizer to track engagement, the whole mechanism passed most of our students by. In this instance it was nothing as complex as extrinsic versus intrinsic motivation. It was a very strategic decision by our students that at the level of the individual only summative assessment items were worthy of attention.

The pedagogical approach described below seeks to use gamification at the level of the cohort, with the intention of motivating students through commitment to their peers not just themselves. Furthermore, the seamless integration with peer instruction in the captive environment of the lecture theatre removes both the apathetic and strategic barriers to participation.

8.4.3 Enhancing peer instruction through gamification

In section 8.1 we explained that our students do not identify as introductory chemists. Instead, there are distinct sub-cohorts consisting of students doing the constituent degree foundation years. It would have been technically possible to subdivide the cohort at the level of the degree programme but given the desire for a critical mass of engaging students, we settled upon three groups: biochemistry and biology; pharmacy and pharmacology; and physical science (chemistry, physics, engineering and mathematics). In the 21/22 enrolment, these three groups were approximately equally sized. The audience response demographic tool allows students to self-identify as belonging to one of these three sub-cohorts without compromising their individual anonymity.

Figure 8.4 depicts modifying the classic peer instruction cycle of figure 8.2 and section 8.2.5 through the differentiation of individual sub-cohorts. The results of the first round of questions can then be presented, sub-divided by their respective group. For competition between the respective groups to promote peer discussion this presupposes that the students are sitting in close proximity. A few weeks into the module, a quick show of hands indicates that this is indeed the case and that our students are arranged in common degree programme clusters around the lecture theatre. This occurs quite naturally without any social engineering on our part.

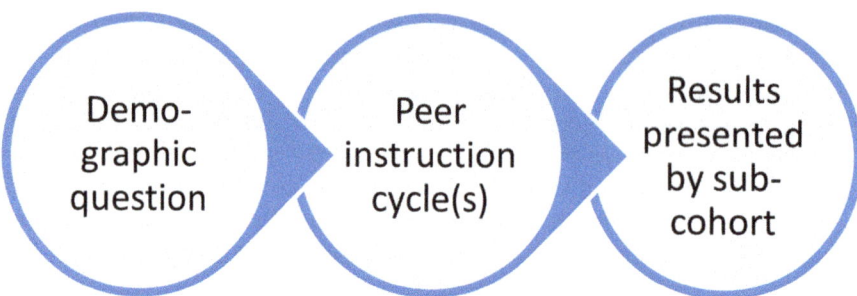

Figure 8.4. Gamification of the peer instruction process.

One of the greatest advantages of peer instruction over other active learning pedagogies is its scalability. An approach that relies on near neighbour interactions to allow a sub-cohort to work collaboratively would appear to be somewhat limited in range. We have overcome this using the Padlet tool first introduced in section 8.4.1.

A Padlet page was established for every demographic sub-cohort. The page was pre-populated with a copy of the peer instruction question. The location of the page is shared with the respective group using a QR code and we use the same page for every teaching session. The combination of 'captive' synchronous communities and incentivization through competition has led to some of the most vigorous discussion we have seen on an online forum in conjunction with Introductory Chemistry. There is nothing stopping inter-group espionage, with students from one sub-cohort choosing to access the Padlet discussion of a second or even third group. From a pedagogical perspective, exposure to multiple conceptual arguments is likely to be beneficial.

8.5 Concluding remarks and advice

We agree with the consensus view that teaching large STEM classes is more effective with an active learning approach. However, we need to be conscious of the diverse experiences, circumstances and aspirations of the students who compose these large classes. Peer instruction is best implemented in-person, even a classically configured auditorium lecture theatre is a much more conducive environment to facilitate and monitor peer-to-peer interactions in large groups of students than video-conferencing software. Less experienced students will only benefit from enhanced learning gains if they are engaged by their teaching. An ideological adherence to a single rigid pedagogical approach will rarely be the best way to serve the needs of teacher and class.

The approach described here includes aspects of both intrinsic and extrinsic motivation. At its heart it seeks to convince students of the benefits of active learning and to convince them to expend just a little more immediate effort in pursuit of lasting learning dividends. What is difficult to convey in a case study is the rhetorical effort involved in encouraging participation. The layering of technologically enabled gamification is a marginal attempt to retain and enhance engagement.

The approach and the subsequent modifications described in this chapter represent a series of compromises between the application of evidence-based pedagogy and the realities of a foundation year course providing service chemistry. While demonstrably more effective (as determined by our application of Lowery Bretz's Bonding Concept Inventory (2014)), a previous fully flipped approach to peer instruction proved less popular with our students, predominantly because it was perceived as more work. Chemistry is being taught in parallel with other science disciplines and we need to strike a balance between the academic foundations and students' attitude towards their studies. Similarly, dedicating two weeks to examination preparation and technique is excessive. However, requests for examination practice begin in week 1 and the notions of teaching and examination preparation are thoroughly conflated in the perceptions of the students and their confidence in the programme.

Since peer instruction is principally a teaching and not an assessment approach, we posit that questions should be primarily judged on how effectively they promote conceptual discussions. In continuing professional development studies, the nature of the question, itself often becomes the focus of the discussion. Thus, the question above about the oak, pivots to a discussion of dried dead wood versus living oak tree. We are confident that those arguing these points later reflect on the impact of the pedagogy. Part of our reluctance to adopt the Mazurian term *ConcepTest* for peer instruction questions is the recognition that they do not need to be subjected to the rigorous validation and reliability testing of a concept inventory to fulfil their purpose.

8.6 Future direction

The teaching practice outlined in this case study continues to evolve. Despite our best efforts to convince our students of the value of formative assessment, as part of the learning process, we remain disappointed with the level of engagement in the formative tests. As a result, the next modification of our approach will see a small allocation of summative marks to each of the end-of-week tests. Furthermore, we will also make the workshops compulsory and stress their role in equipping students to answer the summative assessment questions We accept that this is an admission of our partial failure to convince students of the intrinsic value of the pedagogical approach but believe it is the right approach for this point in the development of English higher education.

Acknowledgements

We thank Grace Arnold for her contributions to using Padlet to foster inter-cohort competition. SJL, the interloper, happily acknowledges his debt to the extended Physics Education Research community for their welcome, support, and inspiration. He would like to particularly thank Dr Anna Wood for her invitation to participate in this book and her wisdom and patience during the editing process. Dr Ross Galloway and Dr Pamela Docherty were generous with previews of their chapters and aid preparing this one.

References

Anderson M, Agger J, Ashworth S, Lancaster S and O'Malley P 2015 Massive open online chemistry *Educ. Chem.* **22** 14–7

Bergmann J and Sams A 2012 *Flip Your Classroom: Reach Every Student in Every Class Every Day 120–190* (Washington, DC: International Society for Technology in Education)

Chamberlain S, Elford D, Lancaster S J and Silve F 2021 Tailored blended learning for foundation year *Chem. Stud., Chim.* **75** 18–26

Dunning D, Johnson K, Ehrlinger J and Kruger J 2003 Why people fail to recognize their own incompetence *Curr. Dir. Psychol. Sci.* **12** 83–7

Lancaster S J, Cook D F and Massingberd-Mundy W J 2019 Peer instruction as a flexible, scalable, active learning approach in higher education *Teaching Chemistry in Higher Education: A Festschrift in Honour of Professor Tina Overton* ed M K Seery and C McDonnell (Dublin: Creathach Press) pp 89–104

Lowery Bretz S 2014 Designing assessment tools to measure students' conceptual knowledge of chemistry *Tools of Chemistry Education Research 155–168* ACS Symp. Series 1166 (Washington, DC: American Chemical Society)

Mazur E 1999 *Peer Instruction: A User's Manual* (Upper Saddle River, NJ: AAPT: Prentice-Hall)

Newbury P 2013 *Writing good peer instruction questions* https://slideshare.net/peternewbury/writing-good (accessed 6 January 2023)

Schell J A and Butler A C 2018 Insights from the science of learning can inform evidence-based implementation of peer Instruction *Front. Educ.* **3** 33

Seery M K 2015 Flipped learning in higher education chemistry: emerging trends and potential directions *Chem. Educ. Res. Prac.* **16** 758–68

IOP Publishing

Effective Teaching in Large STEM Classes

Anna K Wood

Chapter 9

Case study 3: an introductory linear algebra course

Pamela Docherty

In this chapter I describe the structure and design of Introduction to Linear Algebra (ILA) which is taught using active learning techniques. The course consists of ten weekly units, each of which has an overarching learning outcome, and weekly activities. The weekly activities consist of: assigned pre-class activities (reading and a short online quiz); active learning lectures, predominantly using peer instruction; small group workshops; and automatically- and manually-assessed assignments. I also reflect on what has and hasn't worked in terms of large class active learning in our context.

9.1 Introduction

9.1.1 Context

Introduction to Linear Algebra (ILA) is a first-semester, first-year mathematics course at the University of Edinburgh. Students on mathematics and computer science degree programmes are required to take ILA, and students on many other degree programmes may choose it as an optional course. Students on this course come from many countries and have a wide variety of mathematical backgrounds. In particular, some will not have studied Advanced Higher Mathematics, A-Level Further Mathematics or equivalent. ILA is a 20-credit course, equating to 200 hours of study as per the Scottish Credit and Qualifications Framework (SCQF). Students take 60 credits per semester, so they are typically studying two other courses alongside ILA. Depending on their entry qualifications and their performance in a diagnostic test [6], students may be advised to choose as one of their options the foundational, fully-online course Fundamentals of Algebra and Calculus [2]. Otherwise, they may choose an introductory data science course or non-mathematics courses. Since its creation in 2011, the cohort size has ranged from 300 to 900 students.

Due to the large student numbers enrolled on the course, ILA is delivered by a team: a course organiser (the main lecturer who has overall responsibility for the course), one or two additional lecturers, a course administrator, and a team of up to 30 teaching assistants. The three lecturers share the teaching throughout the semester, so each have responsibility for 3–4 weeks of the course content. I was one of the supporting lecturers for the course from 2015 to 2019 before moving to a new role at Heriot-Watt University.

9.1.2 History

In 2011 the mathematics degree programmes underwent a curriculum review which resulted in a re-design of the first year of the programme and brought in a set of new courses. These were ILA; a calculus course (CAP) and an introduction to proofs course (PPS). This chapter will focus on the course design of ILA, but a similar course design was implemented concurrently in CAP and PPS. Around the time of the curriculum review, there was growing interest in active learning techniques in the neighbouring School of Physics and Astronomy (see chapter 7), particularly the ideas of flipped classroom and peer instruction. Colleagues in the School of Mathematics were keen to investigate how these techniques could be implemented in mathematics courses. All three of the new courses implemented a flipped classroom pedagogy, which remains in place at the time of writing (2022). At the same time, the assessment structure and principles were also changed; in particular open-book final exams were introduced.

9.2 Course structure

9.2.1 Overview

The course is structured in ten weekly units, with one topic being the primary focus of the week's study. The weekly activities can be broken down into the following components: assigned pre-class activities; active learning lectures (three 50 min lectures per week); small group workshop (one 80 min workshop per week) and regular assignments, each of which are detailed below. The course follows a standard introductory linear algebra textbook: from 2011 to 2019 the students were required to purchase their own copy of the set textbook [8].

9.2.2 Pre-class activities

Before each class, the students are required to independently read one or two sections of the textbook, then complete an online 'reading quiz'. The reading quiz is an automatically assessed short quiz (2–3 questions) using the STACK [10] assessment system. The goals of the quizzes are: to help students check their understanding of the topics they have just read about; and to give some encouragement for them to complete the reading by awarding some marks (personally, I have since become wary of this approach and no longer award marks for completing pre-class activities). The reading quizzes have a time limit of 25 minutes and students may only attempt the quiz once. Correct answers, marks, and feedback are automatically published after the quiz deadline has passed. The question 'What did you find difficult or interesting in this section?' always appears as the final question in each reading quiz. Responses to this question helps the teacher decide which topics to focus on in the lecture, and this is communicated to students.

9.2.3 Active learning lectures

The lectures take place in a tiered lecture theatre, of varying size between 150 and 400 capacity (in recent years it has been necessary to deliver the lectures twice due to the cohort size).

9.2.3.1 Peer instruction
The course closely follows the standard peer instruction model, the steps of which are described in chapter 2. The majority of lecture time was spent on peer instruction cycles (see chapter 4 for an analysis of how lecture time is spent in ILA and a comparison to some other courses). Generally, we found that we could complete 4–7 peer instruction cycles in one 50-minute class. Marks are never given for correct responses to these polls; we emphasize to students that the goal of the exercise is eventual understanding, not getting it right first time.

As the content of the lecture was not preplanned in the traditional sense, there were no prepared lecture notes. Instead, the teacher wrote minimal notes: a very brief summary or reminder of a key definition or theorem from the reading to serve as a warm-up, then for each peer instruction question, they wrote the question, the correct answer and a brief explanation. The written material that the teacher had generated during the session was uploaded to the VLE afterwards.

9.2.3.2 Peer instruction questions
We found that asking the 'right' kind of questions plays a large part in the success or otherwise of the peer instruction process. Discovering what constitutes the 'right' kind of question was a process of trial and error over many years, and we are undertaking ongoing work to conduct a full analysis of our question bank in order to publish the bank as an open-access teaching resource. In short, the question should probe conceptual understanding of the topic, rather than require a routine implementation of a technique or algorithm. A trio of questions relating to linear combinations and linear independence (concepts which cause much difficulty for students) illustrate this idea:

1. Let $\mathbf{v}_1, \mathbf{v}_2, \mathbf{v}_3, \mathbf{v}_4, \mathbf{v}_5$ be vectors in $\mathbb{R}^n$. Is $\mathbf{v}_1 + \mathbf{v}_2$ a linear combination of the vectors $\mathbf{v}_1, \mathbf{v}_2, \mathbf{v}_3, \mathbf{v}_4, \mathbf{v}_5$?
2. Let $\mathbf{v}_1, \mathbf{v}_2, \mathbf{v}_3, \mathbf{v}_4, \mathbf{v}_5$ be vectors in $\mathbb{R}^n$. Is $\mathbf{v}_2$ a linear combination of the vectors $\mathbf{v}_1, \mathbf{v}_2, \mathbf{v}_3, \mathbf{v}_4, \mathbf{v}_5$?
3. Let $\mathbf{v}_1, \mathbf{v}_2, \mathbf{v}_3, \mathbf{v}_4, \mathbf{v}_5$ be vectors in $\mathbb{R}^n$. Is the zero vector $\mathbf{0}$ a linear combination of the vectors $\mathbf{v}_1, \mathbf{v}_2, \mathbf{v}_3, \mathbf{v}_4, \mathbf{v}_5$?

The options given for both of the above questions are (1) yes, always, (2) never and (3) it depends on the vectors chosen. The first question is often answered correctly by the majority of students the first time as the sum $\mathbf{v}_1 + \mathbf{v}_2$ closely resembles their idea of what a linear combination of vectors should look like. However, it is not obvious to them that $\mathbf{v}_2$ can be expressed as $0 \cdot \mathbf{v}_1 + 1 \cdot \mathbf{v}_2 + 0 \cdot \mathbf{v}_3 + 0 \cdot \mathbf{v}_4 + 0 \cdot \mathbf{v}_5$, and similarly, $\mathbf{0}$ can be expressed as $0 \cdot \mathbf{v}_1 + 0 \cdot \mathbf{v}_2 + 0 \cdot \mathbf{v}_3 + 0 \cdot \mathbf{v}_4 + 0 \cdot \mathbf{v}_5$, so $\mathbf{v}_2$ and $\mathbf{0}$ are also linear combinations of the given set of vectors. The above examples also serve to highlight a focus on abstraction: a concrete linear combination question such as

Is [1 0 1] a linear combination of [1 0 0] and [0 0 1]?

simply encourages use of instrumental knowledge and is unlikely to spark debate.

We found the best questions tend to have a correct response rate of 50%–80% on the first vote: any less than 50% showed that the majority of the class was confused and would struggle to arrive at the correct answer without further support. Questions with a first vote correct response rate of higher than 80%, on the other hand, can be used in moderation to elicit feelings of achievement in the class, but are not worthwhile as a peer instruction activity. The question bank contains 5–10 questions on each topic: the teacher might have a rough plan for which questions they are going to ask in the lecture, but can be flexible, skipping questions and choosing to ask different questions off the cuff depending on the class voting results. The questions on a topic often follow a progression of increasing difficulty, allowing a scaffolded approach to understanding.

In principle, student responses to the 'what did you find most difficult?' question in the pre-class reading quiz informs the design and focus of the lecture. In reality, the course team already has a very good understanding of which concepts students find the most difficult, informed by their past teaching experience and the literature (see e.g., [1, 9]). The responses to the question 'what did you find most difficult?' are much the same from year to year.

9.2.3.3 Electronic voting system

For the purposes of polling, originally a voting response system ('clickers') was used, which was eventually replaced by an electronic voting system (EVS) (TopHat[1]) which students access via their personal web-enabled device (phone, laptop, or tablet). This particular EVS has many question types, such as word answer, numeric answer and click on target questions, although we use the multiple choice option (MCQ) the most frequently. There are a couple of reasons for using MCQs the majority of the time: the multiple choice option allows the best side-by-side visualization from the polling results of the first and second votes, so the students can see the switch from incorrect to correct answer most clearly. This demonstrates an important consideration in the success or otherwise of the peer instruction

---

[1] https://tophat.com/.

technique: students need to see that it is working. We also found that free-text responses were to be used with caution, as the responses will be displayed to the class. We did not ever experience any offensive comments, but we experienced plenty of (deliberately, we assume!) comically incorrect numeric answers. Seeing these gives the class a laugh for a few seconds, which sometimes can be valuable to lighten the mood, but we've found that it can throw off the pace of the lecture, and it can be hard to get back on track.

9.2.3.4 Classroom management
The students were not assigned groups to sit with in class. This is different from the Mazur-proposed model of peer instruction, where it is advised that students sit with the same group each class. The ILA course team did not assign students to groups for practical purposes: it is hard to imagine the practicalities of 300 students entering a lecture theatre and sitting in an allocated seat. However, we appreciate that there is likely to be added benefit from working with the same group every time. We simply asked students to turn to those sitting around them to discuss. This meant they may be discussing with people they have never spoken to before, especially in the early weeks. This was likely a cause of some nervousness to begin with, but students soon realized that group participation was an essential component of class time and would be every time, so they soon knew what to expect. In practice, as the course got underway, students would often form friendship groups and choose to sit with them each lecture.

Each year, we encounter a handful of students who prefer not to participate in group discussions. That is their prerogative, and the course team support them in participating in the class in a way that feels comfortable to them. The teacher may occasionally seek them out during the discussion periods to ask them their thoughts on the question: this is our way of encouraging discussion and giving them tacit support that we are not forcing them to speak to their peers if they are not comfortable doing so.

9.2.4 Workshops

Students are allocated to a workshop group of around 12 students which meet each week for 80 min, supported by a teaching assistant (TA). The goals of the workshop are the following: to get students working on and discussing mathematics together; and to help them develop their mathematical skills in a supportive and informal environment, with regular interaction and feedback from a TA.

9.2.4.1 Teaching studios
We believe that the physical space was crucial to the success of these workshops. The University of Edinburgh has a number of rooms called 'Teaching Studios', the design of which consists of multiple (12–16) semicircular tables which seat five or six. Each table has a PC, together with a large screen, whose feed can be switched between the table PC and what the teacher is displaying from the teaching desk.

Multiple workshop groups (6–8) meet in the same room at the same time, with each TA looking after two tables of students, overseen by a supertutor (usually a senior member of the course team) who opened and closed the sessions. This led to a 'buzzy' atmosphere: there felt like a shared purpose and community spirit to the event, as compared to a more traditional flat classroom setting where one tutorial group is meeting. However, we acknowledge that this environment does not represent ideal learning conditions for many people, particularly those with neurodivergent profiles: some workshop groups were arranged in a smaller, quieter teaching space and students with certain learning profiles were offered the option of attending these groups.

The shape of the tables seemed to have a positive effect to encourage collaborative work: the students are facing each other, and there is just enough table space in the middle for them to all easily see and reach some shared work.

To further encourage collaboration, each group was provided with mini (A3) tabletop whiteboards to encourage shared written work and discourage students from defaulting to writing their own solutions on their own paper and not discussing with the others.

The TA is also responsible for marking the weekly homework of the students in their group (see next section). The TA was, therefore, able to give some verbal feedback to the students during the workshop as well as written feedback on their homework, and students could ask any follow-up questions there and then. The weekly workshop with a fixed group and fixed TA for the duration of the course seemed to foster a sense of community, which encouraged attendance: students felt more accountable as their presence or absence would be noticed[2].

9.2.4.2 Preparation or not?

Over the years, the course team experimented with whether or not to request that the students prepare the questions in advance. When the questions are released in advance, some students will complete them all and some students will come unprepared, leaving the TAs struggling to engage the students in meaningful work. When the questions are not released in advance, all students come to the questions 'cold', so are more likely to engage in discussion with their group. However, some students would prefer to have a chance to prepare the questions, as they feel they are slower than their peers and want to have a chance to prepare. We have not found a good solution to this conundrum.

9.2.5 Assessment structure

The assessment structure is composed of a continuous assessment component (20%) and a final exam (80%). This course makes extensive use of the STACK assessment system (see chapter 6 on assessment methods), along with many other courses in the School of Mathematics at the University of Edinburgh.

---

[2] Historically, workshop attendance rates were reasonably high although decreased as the semester wore on. Post-pandemic, attendance is much lower and this is an issue the course team is trying to overcome.

The continuous assessment component is made up of three components:
1. reading quizzes (5%)
2. in-class engagement (5%)
3. weekly hand-ins (10%)

9.2.5.1 Reading quizzes
The reading quizzes were already described in section 9.2.2. There are two reading quizzes each week, roughly corresponding to 2–3 sections of the textbook which would be covered in the subsequent lectures. There were roughly 20 quizzes in total, each worth 2–3 marks, and the best 16 out of 20 quiz scores constituted 5% of the students' final grade.

9.2.5.2 Weekly homework
A selection of exercises from the textbook was assigned as the weekly homework. The purpose of these assignments is for the students to practice writing solutions professionally, receiving regular feedback on the quality of their mathematical writing from the same tutor. The solutions were handwritten, handed in and marked by a teaching assistant. Originally, there were roughly 10 weekly homeworks in total, worth 10 marks each, and the best 8 out of 10 marks constituted 10% of the students' final grade. As the cohort size grew, the number of written homeworks reduced from 10 to around four, and further STACK assessments were implemented to compensate for the others.

9.2.5.3 In-class engagement
The final 5% of the continuous assessment grade component was essentially an award for in-class engagement. This data was collected through the EVS: when a student voted in at least two peer instruction questions in a given lecture, they were recorded as having attended and engaged with the lecture. Students were not expected to attend every lecture: students would receive the full 5 percentage points if they had attended around 80% of the lectures. On the other hand, if they had attended less than half of the lectures, they would receive 0.

9.2.5.4 Final exam
One of the most important features of ILA and the other courses introduced in the 2011 curriculum review was the introduction of open-book final exams. At the time, this was extremely uncommon in UK mathematics degrees [3, 4], and is still not the norm today [5]. By open-book exam, we mean that students could bring in the following specified items: a copy of the set textbook (and no other bound texts); an unlimited amount of their own printed or handwritten notes. The format of the exam gives dual consideration to assessing both conceptual understanding (and associated reasoning skills; a typical question might ask whether a given statement is true or false and to justify their choice with a proof or counterexample) and ability to carry out standard linear-algebraic procedures, such as Gaussian elimination or diagonalization of matrices.

9.3 Evaluation

The course team has not undertaken a formal evaluation of the success of the course design similar to that described by Ross Galloway in chapter 7. There isn't a well-validated tool akin to the Force-Concept Inventory for Linear Algebra that could be used. If we wished to consider the impact on student grades as a proxy for student learning (which is well understood to be problematic in itself), this would be impossible because the introduction of more active learning techniques in the classes coincided with a curriculum review and the introduction of entirely new courses, so course grade distributions cannot therefore be compared to previous instances of the same course. Anecdotally, however, it is generally agreed that the levels of achievement in terms of course grade distribution and achievement in follow-on courses is at least as good as it was previously, and most likely better.

It is, however, abundantly clear that the implementation of more active learning techniques in this and other courses has led to a marked improvement in the feeling of community amongst students and staff in the School of Mathematics. The School is often rated highly for learning community in the National Student Survey, and in 2017 was rated top amongst Russell Group mathematics departments in this category. We are certain that the design of the first year courses to foster interactive engagement has been a large contributing factor. Other contributing factors to the learning community not so far mentioned are: the widespread use of the online Q and A software Piazza[3]; a successful Peer-Assisted Learning Scheme (MathPals) and a widely-used drop-in helpdesk (MathsBase).

As mentioned previously, and again similarly to the physics course described in chapter 7, we have a large amount of data of the polling results of every peer instruction question posed in class over five years. We are currently undertaking research on this to figure out what makes a good quality peer instruction question, and we plan to publish the question bank along with evaluations as an open-access teaching resource.

9.3.1 What didn't work, changes post-COVID-19

It's important to note that we haven't achieved the 'perfect' course design. We have tried many things over the years that were quickly abandoned, and the course team still struggle with some aspects.

There were some major changes made to the course as a result of COVID-19. For example, the set textbook is now an open-access textbook which is freely available online [7], and students now submit their weekly homework online, where they are marked electronically using Gradescope[4]. Some lecture sessions were replaced by online 'lecture quizzes' which constitute short videos interspersed with interactive quizzes in STACK, which students are required to work through as part of their independent study. The interactive quizzes encourage the students to be active in their independent study; but the sense of community and value of the peer instruction process is lost. Since 2022, the final exam has been almost entirely

---

[3] https://piazza.com/.
[4] https://www.gradescope.eu/.

automatically assessed [11]. At the time of writing, it is unclear which changes are to be made permanent.

Acknowledgements

I wish to express my sincere gratitude to Chris Sangwin, Sue Sierra, Tadahiro Oh, James Lucietti, Carlos Zapata Carratala, Louise Durie and many others who have been part of the ILA teaching team over the course of the past decade. Thanks in particular to Toby Bailey, who as Director of Teaching and Course Organizer for ILA was instrumental in making the enhancements described in the chapter, in which I played a minor part. I'm grateful for the opportunity to document this on behalf of all my ILA colleagues.

References

[1] Dorier J 2000 *On the Teaching of Linear Algebra* (Dordrecht: Springer)
[2] Gratwick R, Kinnear G and Wood A K 2020 An online course promoting wider access to university mathematics *Proceedings of the British Society for Research into Learning Mathematics* vol 40 1st edn ed R Marks (Edinburgh: The University of Edinburgh)
[3] Iannone P and Simpson A 2011 The summative assessment diet: how we assess in mathematics degrees *Teach. Math. Appl.* **30** 186–96
[4] Iannone P and Simpson A (eds) 2012 *Mapping University Mathematics Assessment Practices* (Norwich: University of East Anglia)
[5] Iannone P and Simpson A 2022 How we assess mathematics degrees: the summative assessment diet a decade on *Teach. Math. Appl.* **41** 22–31
[6] Akveld M and Kinnear G 2023 Improving mathematics diagnostic tests using item analysis *Int. J. Math. Educ. Sci. Technol.* 1–28
[7] Nicholson 2019 *Linear Algebra with Applications* (Calgary: Lyryx)
[8] Poole D and Lipsett R 2014 *Linear Algebra: A Modern Introduction* (London: Cengage Learning)
[9] Rensaa R J, Hogstad N M and Monaghan J 2020 Perspectives and reflections on teaching linear algebra *Teach. Math. Appl.* **39** 296–309
[10] Sangwin C 2013 *Computer Aided Assessment of Mathematics* (Oxford: Oxford University Press)
[11] Sangwin C 2019 Developing and evaluating an online linear algebra examination for university mathematics *Eleventh Congress of the European Society for Research in Mathematics Education* (No. 15) (Utrecht: Freudenthal Group; Freudenthal Institute; ERME)

Chapter 10

Case study 4: personalised learning by student-posed questions during biology lectures

Heather McQueen

Quectures are flipped lectures that employ student-posed questions as an active learning technique. These student-posed quecture questions personalise student learning in lectures and raise student attainment, particularly for students that were previously poorly engaged and for low scoring students. This chapter describes the motive and method for introducing this technique into a large undergraduate biology class. It will then discuss benefits and issues that were experienced or evidenced while using the technique. This will be followed by a discussion of the quecture technique in an online learning context and some basic rules for introducing quecture questions into your own teaching.

10.1 Are students talking in your class? Should they be?

Scenario one: You stand at the front of the lecture theatre, mid-way through a rich and contextualized explanation of one aspect of your lecture topic. The lecture theatre is in complete silence except for your voice. You pause and look meaningfully at your audience to allow your interesting point to settle and ask if there are any questions. Most of the students in the front row stare blankly at you while a few in the rows behind are sleeping. Someone raises their hand. 'Yes' you say eagerly. The student says 'Do we need to know this for the exam?'.

Scenario two: You are mid-way through executing the plan for your interactive lecture. There is no-one on the stage and the class is alive with chatter. No-one is talking about exactly the same thing, because each student is engaged in a discussion about the aspect of the lecture topic that they, or their classmate, most wish to better understand and you are amongst the students, simply listening. How are you ever going to get to the end of your plan with so many questions?

Which scenario do you prefer and why? From a pedagogical viewpoint, the main differences are arguably: (i) who is talking; (ii) who is doing the hard thinking; and (iii) the nature and level of learning that is taking place. In scenario two the student is likely to be doing all three while in scenario one many students will be doing none of these things and no-one will be doing all three except, perhaps, the teacher.

Having experienced too many scenario one lectures I decided to flip my classroom (described in chapter 2), and I wanted to be sure that this would really encourage improved engagement and learning, not just for the more successful students but for all my students. This class met in our largest lecture room, which was also the usual venue for our biggest guest seminars. When I compared those sleepy, blank, student faces with the faces of staff sitting in exactly the same seats and for the same reason (to learn), I realised what it was that I wanted for my flipped classroom. I wanted students that were engaging critically, just like the guest seminar audience, eagerly trying to connect this new knowledge with known concepts, looking for flaws or gaps in the argument (or their understanding of it) and working to grasp exactly what consequences this could have for their own work or learning. Of course, some students already approach their studies with this academic intent, but many students, particularly those who are inexperienced or have prior educational disadvantage, have no experience of this approach to learning and would need to be taught how to do this.

Luckily for me, at the time of this epiphany I was enjoying a secondment with the Institute of Academic Development at the University of Edinburgh, which provided me with time to read and reflect, and supportive knowledgeable colleagues with whom to discuss ideas for my flipped classroom. (I thoroughly recommend a pedagogical secondment to any busy academic who cares about their teaching. It was a career-changer for me.) One pivotal moment was discovering the quote 'Thinking is driven by questions. Answers are a full stop to thought' (Foundation for Critical Thinking 2019). This was it! I wanted my students to ask questions during lectures, and to think about them and to talk about them rather than to receive answers. And so the 'quecture' was born.

10.2 How can we encourage personally relevant learning for every student during large class teaching?

Learning to construct their own questions is known to lead to gains in student comprehension (Rosenshine et al 1996) yet the technique of student-posed questions is under-valued and under-used in the science classroom (Chin and Osborne 2008). Quectures are my version of the flipped classroom that incorporates the in-class activity of students identifying and engaging with their own questions (McQueen and McMillan 2020). Supporting this activity during learning in the lecture theatre aims to encourage struggling or previously unengaged students to explore their misconceptions or knowledge gaps around even the most basic concepts or principles. Misconceptions are known to be important to address since they are often particularly resistant to change and can block new learning if things don't 'add up' or agree with what the student already believes that they 'know'. For a review on changing student misconceptions see Taylor and Kowalski (2014). While a struggling student will benefit from addressing their area of personal difficulty within new learned material, at the same time a flourishing student will engage by posing advanced knowledge-synthesis, or more esoteric 'wonderment' questions (Wood et al 2018) thus personalising their learning.

So how can such engagement with student-posed questions (quecture questions) be incorporated into a lecture and how can we ensure that the learning will be relevant? The trick is to build the activity around those all-important lecture learning objectives. You cannot spend too much time making sure that your learning objectives are appropriate. They are your pact of common understanding with your students, making clear what they should be learning and how they should use that knowledge. They set the learning content, the teaching plan and the assessments, and they are my go-to answer if ever I get that 'need-to-know-this' question. 'You need to know whatever will help you to achieve the learning objectives' immediately bats the responsibility for learning back to the student. I consider that less is better here and normally aim for one or two learning objectives per lecture, or three at most. My learning objectives are always phrased as something that the students should be able to do as a result of the lecture. Bloom's verbs (Anderson and Krathwohl 2001) provide a good list of suggestions for writing these. For a guide to writing and using learning objectives see Orr et al (2022).

Lecture content for each quecture learning objective is provided as a 'to-do' list that students work through in preparation for the quecture (flipped lecture). This preparation includes advice to 'consider anything that you find difficult or that raises further questions for you' and to 'construct your own questions' relevant to each learning objective. Students considering their own question during the preparation phase is of critical importance since identifying and articulating your own most pertinent question is cognitively demanding and could prove too difficult to complete if only begun during the more demanding environment of the live lecture.

During each quecture the learning objectives are addressed in turn via teacher explanation and student interaction, often via peer interaction as described in chapter 2. After each learning objective has been covered there is a three-minute pause for the students' own quecture questions. Using visual and verbal prompts students are instructed to 'think, type and talk', in that order (figure 10.1). The thinking is important as described above, and because a student's question may well have developed and require re-structuring, or replacing, as a result of in-class learning. It is then important that the student commits to their question by clearly articulating it and writing or typing it. We ask students to submit their question using their mobile device logged onto our personal response system for ease of collection, but writing on paper would work just as well. Submitted questions, which are anonymous to the rest of the class but not to the teacher, are seen by the whole class who can respond or simply 'like' a question which interests them. In the third step of the sequence students are encouraged to turn to nearby students

Learning objective 1:

Recognise and compare modes of single gene inheritance involving autosomal or sex chromosome genes, and dominant or recessive phenotypes.

Figure 10.1. Example slide used as a visual prompt to encourage students to engage with their own quecture question during lectures. The same prompt with substituted learning objective is used for each quecture question.

and discuss their own question. This introduces a social element which, on its own, is a good thing for any large class, but also allows some brief collaborative learning which is also of great value, particularly for students without friends or family with academic experience. Discussion of questions allows students to challenge each other's ideas, construct explanations together and better understand key concepts (Chin and Osborne 2008).

As with any new teaching technique, it is of critical importance to convey the purpose and method of the quecture question before commencing. Explanatory text and videos can be included in the course book, the virtual learning environment, introductory lectures and any other student communications to convey to the students what they should expect to gain from engaging with quectures (which can be summarised as 'to retain more and apply better'), and how to do so. We illustrate four main example question example types for students: (1) is something missing from my knowledge?; (2) is something contradictory?; (3) does something link with something else?; or (4) a deeper 'wondering' question such as 'how do we know this?'. The two simple rules for quecture questions are that each question a student constructs is: (a) relevant to the learning objective, and (b) for themselves to answer.

Many of the students' questions will not be answered by the brief peer discussion in class and, although the last step in the quecture strategy involves the teacher revisiting submitted questions at a specified future lecture, even then the mission is not to provide answers. Remember that 'full stop to thought'? Questions are re-visited at a future lecture partly to employ the distributed practice effect (Dempster 1989) (i.e., learning is improved when repeated at separate sessions rather than considering an

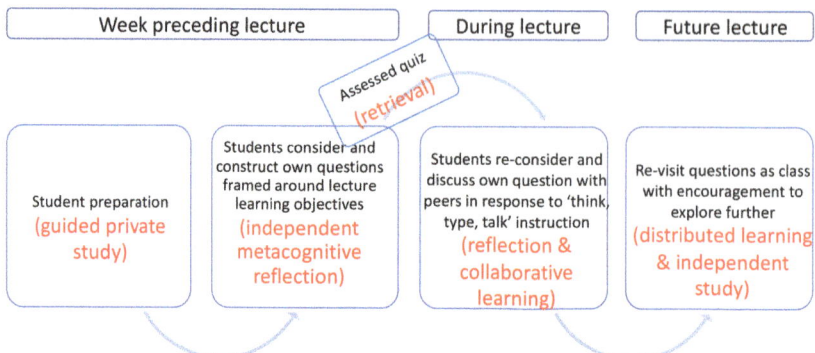

Figure 10.2. Schematic of learning activities and methods throughout the stages of the quecture strategy. Student learning methods at each stage are in red.

idea at one session only), but also to provide suggestions and encouragement to students to continue to explore their own questions. The purpose of the quecture question is not to elicit the teacher's answer, but is instead to improve student learning habits by practising reflection, self-questioning, discussion and independent enquiry. The overall strategy with respect to what the student is doing pedagogically is illustrated in figure 10.2. You can also see a detailed analysis of activities and timings captured during live quectures using the FILL tool in chapter 4 (figure 4.2).

10.3 How was the quecture strategy received?

From a teacher perspective, implementing my first ever quecture session was both scary and invigorating. I will never forget my own uncomfortable reflections on how unusual it was to ask students to do this during that first long silence in a room with over 300 students. Maybe too unusual? Would the students play along or would they just use the opportunity to catch up on the student chat? It probably took more than a minute of relative silence, one of the longest minutes of my life, before the first student question appeared on the student response system. In hindsight, of course, this was as expected as the students needed this time to identify, formulate and compose their question. With immense relief I realized that the first submitted text was a real and relevant question. Then another, then another, then an avalanche. More questions than I could ever answer, but then it was never the plan for me to address them.

It is hard, sometimes impossible, to resist the urge to answer a student question, especially given that you are their teacher and an answer is generally expected. But remember that a student asking a question now has some investment in that question and if you provide an immediate answer you will remove, forever, their opportunity to discover the answer for themselves, and more importantly, to practise this potentially new-found independent learning approach.

Investigating the student perspective, we found that the strategy was welcomed and appreciated by many students who recognised the intended benefits and acknowledged that this style of learning re-balanced the responsibility for learning during lectures, away from the teacher and towards themselves taking responsibility (McQueen and McMillan 2020). The types of questions students submit are satisfyingly broad and, as you will see from the sample of questions submitted in response to one learning objective, include examples of all four example question types discussed above (table 10.1).

Table 10.1. A range of quecture questions submitted by students in response to the learning objective to be able to 'recognise and compare modes of single gene inheritance involving autosomal or sex chromosome genes, and dominant or recessive phenotypes'. Questions are presented as submitted by students.

What is the difference between x linked recessive and x linked dominance?
Are there any easy ways to determine the difference between sex linked and autosomal?
Does recombination occur between homologous sections of the X and Y chromosomes?
What do you draw on a pedigree if you don't know someone's genotype/phenotype?
What does 15q21 means?
Why are sex-linked traits a thing?
Are there many cases of y linked conditions?
How do we know those human conditions resulted from single gene mutation?
Under what circumstances can a female inherit an x linked recessive affliction?
What distinguishes sex-limited or sex-influenced inheritance from sex-linked inheritance on a pedigree?
What cases result in the fsther giving his autosomal recessive trait to his daughters?
For a male, if X is inherited from the mother, and Y from the father—for a female, how does she become a carrier for a sex linked disease her father has if she doesn't inherit the Y chromosome?
What happens if dad is carrier and mother is heterozugous, is it a 50/50 chance of the phenotype showing?
Is it ever possible for a female individual to only inherit their X chromosomes from their mother?
Why is wild type used instead of just saying 'dominant'?
What do modes of inheritance and pedigrees look like for complex, multi-gene traits?
How do we calculate the probability of a particular genotype in a pedigree?
What conditions if any, involved the phenotype masked by males, so present in females?
Can translocation cause an x linked gene to become autosomal? Or vice versa?
Are there any human diseases which are lethal when homozygous?
How do hormone environments affect sex limited/influenced genes?
How do we distinguish recessive from de novo mutations?
If a gene is on the autosomal region of the Y chromosome, does it follow a sex linked inheritance or autosomal one?
How closely related does a pairing have to be for it to be consanguineous?
How does autosomal polymorphism become so common in normal society?
Why was X chromosome evolved. Does the benefits of sexual reproduction outweigh the risk if sex linked diseases? Is it a result of sexual reproduction?
Are mutations in mitochondrial genes sex-linked?
What is a probality of an x linked disorder in a female, if one of x chromosomes is turned into a Barrs body in each cell?

10.4 Can the quecture strategy really make a difference: potential benefits?

Aside from staff and student feelings and perceptions can the use of quecture questions within lectures improve learning and therefore improve student course scores relative to those students who did not engage with the quecture questions? The answer is yes. Student data shows that engagement with quecture questions is associated with improved course scores, and that at least some of this improvement is specific to the use of quecture questions and not just to strong general engagement with learning (figure 10.3). We also found that the effect of quecture questions is greatest for students with low prior course scores and for those with low general engagement (for more detail see McQueen and Colegrave (2022)). This is exciting because students who do not engage well and who receive low scores are the very learners that are most in need of improvements to their learning methods and to their learning outcomes.

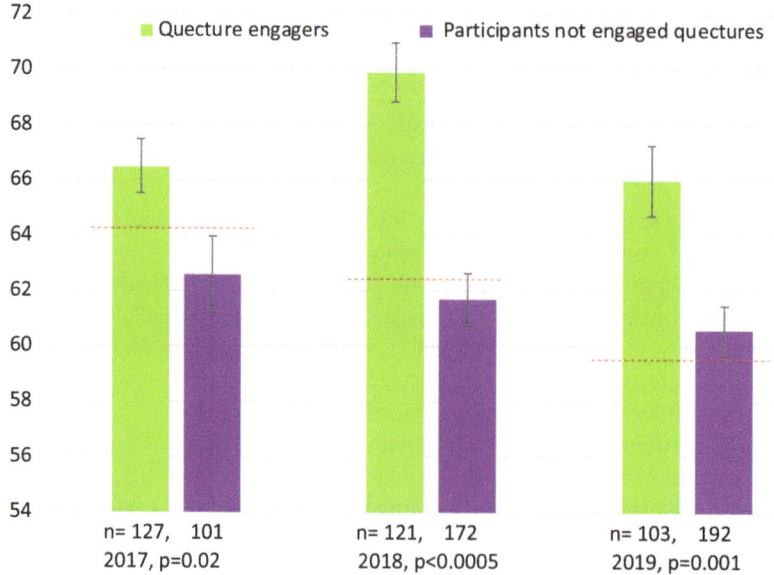

Figure 10.3. Course scores for engaged students that either did or did not submit quecture questions via the personal response system. Average course scores for students who submitted their own quecture questions (quecture engagers) is shown in in green alongside scores for students who participated in other personal response system questions but did not submit quecture questions (participants not engaged quectures) in purple. Three iterations of the same course in 2017–2019 are shown. The number of students whose score has been averaged is shown below each bar and the average course score for each of the three years is shown as a red dotted line. In 2019 the average course score was lower than both populations due to the low marks of students who did not engage using the personal response system.

10.5 Can the quecture strategy really make a difference: potential pitfalls?

It would, however, be misleading to suggest that the introduction of the quecture questions was an unmitigated success as the strategy generated something of a marmite response (some loved it, others hated it), leading to an issue of low and biased student engagement. Submitting a quecture question via the personal response system was a voluntary activity, and our data shows that around one third of students submit questions. Many students may well have considered a question without posting but submission was the only metric of engagement available to us. Moreover, groups of students who traditionally struggle at university (such as first in family to attend university) were under-represented in the group that did engage (McQueen and Colegrave 2022). Ironically, as mentioned above, those groups were also found to enjoy the highest benefit from the strategy when they did engage. While the problem of poor engagement is certainly not unique to the quecture strategy, it is an important one to address if the strategy is to achieve its aim of encouraging personally relevant learning for every student. Arguably extra coaching in the methods and benefits of constructing questions would provide much-needed support for students to develop their reflective metacognitive approach to learning and potentially strengthen engagement with the quecture strategy.

A second potential pitfall was misunderstanding of the purpose (particularly student ownership) of quecture questions. Interestingly, the misunderstanding that the questions were for the teacher to answer (rather than for the student to take ownership of) was as prevalent in teaching staff (not implementing the strategy) as it was in students. It is impossible to be too explicit about the importance of student ownership of questions which is critical for the longer term benefits to students. Without student ownership of questions, the technique would simply be another question-and-answer opportunity.

One mistake that I made, once and only once, was to allow humour into the strategy. One particularly amusing student question about male pattern baldness which took the form 'asking for a friend' had me laughing along with the class. This, however, was quickly followed by a mischievous burst of light-hearted questions that were not relevant to the learning objectives which, sadly, I had to block with a stern warning. Of course, humour is a great tool for engaging, socialising and relaxing any group, and is normally to be welcomed in teaching situations. However, for this new, difficult and slightly awkward task of considering and constructing a genuine learning question, turning to humour proved to be an easy get-out strategy for some students, which won the admiration of the class but spoiled the asker's opportunity to capitalise on these few minutes of metacognitive learning in class.

Lastly, the utmost care and sensitivity is required for question handling. No matter how non-genuine, poorly phrased or unconsidered a question may seem, one should never deride any question, even lightly or in fun. A question may appear at first glance to have been put together with no effort or with disregard for what has just been taught or even with intent to offend, while in reality the question can be a genuine attempt to understand by a thoroughly confused student. For some students, perhaps those most in need of the intervention, this seemingly non-genuine question may well represent a metaphorical dipping of a toe in the water of becoming the sort of student who articulates their own question.

10.6 How can the quecture strategy be exploited when lectures have moved online?

Following the rapid move to online learning across universities in 2020 as a result of the global pandemic, many university teachers have learned for themselves that the flipped model of providing learning material in advance and following up with teacher interaction works. Since the quecture strategy was originally designed as an intervention within live lectures, does it still have validity if this new model is retained and content is all presented online in advance? Yes it does.

Building on pre-pandemic experience of the quecture strategy at this university, a small number of courses at first, second and honours years within the School of Biology adopted a quecture-based structure for online lectures during, and indeed since, the pandemic. Lecture content (as videos or as richer lists incorporating reading and other activities) is provided for asynchronous study prior to a synchronous session with the lecturer, with the final asynchronous student activity being to formulate and post their quecture question on a shared online site (Padlets from padlet.com or virtual learning environment discussion boards). Arguably this asynchronous method of committing to a question represents an improvement on the original in-lecture quecture strategy because it allows the students to fully consider their own questions in their own time, in contrast to doing so during the high cognitive load that some students may experience during a live lecture.

One of our biology courses signalled the value placed on this activity by requiring students to submit their best question at the end of teaching, along with an account of how they had explored possible answers. Students were rewarded for this submission with up to 5% of course assessment points, where marks were awarded for evidence of effort towards addressing the question rather than the level of the question itself. The assessed quecture question submissions made interesting reading, ranging from clarification of basic understandings by some students to answers that demonstrated impressive academic depth, described by the course organiser as 'inspiring'. The course organiser also commented 'I think that it is the best thing we introduced into the course', and '5% of the course assessment well spent'. Personally, I was particularly struck by the large number of questions that pertained to very specific circumstances that were clearly of significant personal interest to the submitting student, indicating the technique to have captured strong personal investment in their learning.

10.7 How can you use quecture questions in your large classes?

Quecture questions are student-posed questions that aim to better engage students with learning content by encouraging personalised construction of knowledge, and to build habits of reflection and discussion for learning. By engaging with their own questions in this way, students can improve their learning and their course scores.

This is particularly effective for students with low prior scores and poor prior engagement. Assuming the benefits of the quecture question to be not limited to biology students, structuring and supporting this activity during either traditional in-person lectures or during asynchronous learning can contribute to improved learning and to improved equity within your classroom.

Although this chapter describes two different defined lecture strategies (in-person quectures and flipped online lectures), there is no reason that quecture question learning benefits should be restricted to these scenarios. Indeed, the quecture question could be employed as a learning intervention in any learning situations where there is a defined learning objective. To do so, the main task for the instructor is to convey clear explanations about the intervention to the students and, as such, this is not a time-consuming proposition for the teacher, but does require allocation of a few minutes for the students to engage during each learning session.

The features essential to success with employing the quecture strategy could be listed as:
- Frame the prompt for questions around a learning objective.
- Allocate time for students to reflect upon their own questions. If being done in a live teaching situation ensure that students have also had time before the live event for this.
- Have your students write down (and thus commit to) their question.
- Encourage student discussion and further independent study of each question.
- Support the students to understand both the purpose and the method of the strategy, ideally including formal instruction and practice with posing good questions.
- Be explicit about student ownership of the questions.
- Employ the strategy with sincerity and genuine care for teaching. If you are doing this half-heartedly to tick some boxes you are wasting everyone's time.
- Be kind!

Scenario three: You are mid-way through your newly re-designed in-person teaching session. The class is alive with chatter. No-one is talking about exactly the same thing, because each student is engaged in a discussion about the aspect of the topic that they, or their classmate, most wish to better understand, and you are amongst the students, simply listening. You pause and smile to yourself. There are so many more questions than you could ever answer. The opportunities for student learning are limitless.

References

Anderson L W and Krathwohl D R 2001 A taxonomy for learning, teaching, and assessing *Abridged Edition* (Boston, MA: Allyn & Bacon)

Chin C and Osborne J 2008 Students' questions: a potential resource for teaching and learning science *Stud. Sci. Educ.* **44** 1–39

Dempster F N 1989 Spacing effects and their implications for theory and practice *Educ. Psychol. Rev.* **1** 309–30

Foundation for Critical Thinking 2019 https://www.criticalthinking.org/pages/the-role-of-socratic-questioning-in-thinking-teaching-amp-learning/522 (accessed 4 May 2022)

McQueen H A and McMillan C 2020 Quectures: personalised constructive learning in lectures *Active Learn. High. Educ.* **21** 217–31

McQueen and Colegrave 2022 Raising attainment for low-scoring students through quectures: an analysis of achievement and engagement with personalised learning in lectures *Int. J. STEM Educ.* **9** 44

Orr R B, Csikari M M, Freeman S and Rodriguez M C 2022 Writing and using learning objectives *CBE—Life Sci. Educ.* **21** 3

Rosenshine B, Meister C and Chapman S 1996 Teaching students to generate questions: a review of the intervention studies *Rev. Educ. Res.* **66** 181–221

Taylor A and Kowalski P 2014 Student misconceptions: where do they come from and what can we do *Applying Science of Learning in Education: Infusing Psychological Science into the Curriculum* ed V Benassi, C Overshon and C Hakala (Washington, DC: American Psychological Association)

Wood A K, Galloway R K, Sinclair C and Hardy J 2018 Teacher-student discourse in active learning lectures: case studies from undergraduate physics *Teaching Higher Educ.* **23** 818–34

Effective Teaching in Large STEM Classes

Anna K Wood

Chapter 11

Case study 5: the learning assistant model for engaging students

Valerie K Otero

Learning Assistants (LAs) are undergraduate students who are hired to participate on the instructional team for university courses. They increase the teacher–student ratio and make it possible for instructors to implement active learning pedagogies. LAs are involved in planning and reflecting with the instructor, providing feedback to students about key concepts, course navigation, and where to get mental health resources. LAs mediate between students and the institution leading to feelings of inclusion and belonging, and to improved learning and retention. The LA model has spread throughout the world and instructors engage with one another through the International LA Alliance.

11.1 Introduction

Active learning classrooms can take many forms, from students discussing with their neighbour in large lectures to flipped classrooms that engage students in small group work facilitated by research-based, open-ended problems and questions. Regardless of the approach that is used, helping students interact with one another is challenging even to the best of instructors, and especially in large-enrollment classrooms where students can easily feel voiceless and underserved.

The Learning Assistant (LA) model has been adopted by universities all over the world as a way to increase the teacher to student ratio in the college classroom through the inclusion of specially trained undergraduate students who are hired to serve on the instructional team of college courses. While students can apply to be an LA for all approved course, they usually apply to serve as an LA for courses they have already taken. Similarly, instructors typically select students as LAs if they have already taken the course, although this is not always the case, especially for courses that are new to using LAs. LAs assist faculty in the modification and enactment of instructional innovation. In addition to helping enact active learning

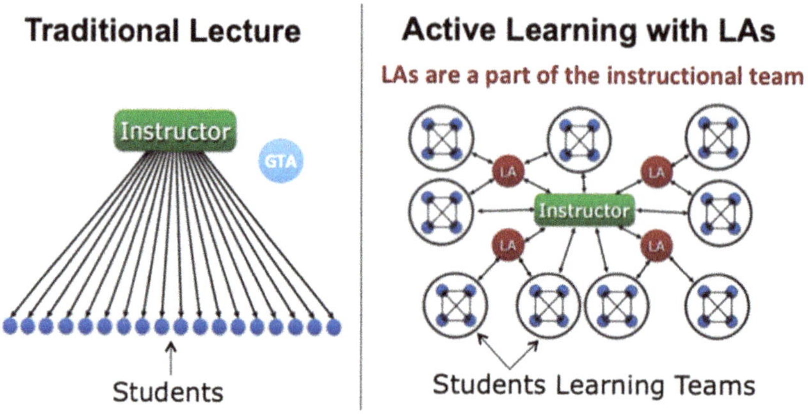

Figure 11.1. Traditional lecture compared to LA-supported course. Undergraduate LAs serve on the instructional team and interact with student groups during class time.

pedagogies in the classroom, LAs provide relevant guidance to students on everything from how to study for the course to where to find mental health resources on campus (figure 11.1).

Faculty typically become interested in working with LAs when they seek to make changes to their courses for a variety of reasons, for example, to improve learning outcomes for a greater diversity of students, to reduce failure rates, to engage students actively during classroom time, and to improve students' sense of inclusion and belonging in their course and discipline. LAs help faculty members make changes to their courses and provide them with important insight about how students are experiencing the course.

LAs engage in three main activities, they: (a) meet weekly with the lead faculty member of the course to co-plan and prepare for the upcoming week, and reflect on the previous week, (b) lead learning teams in classrooms, where students work in small groups on group-worthy activities and LAs listen to group conversations and question groups to help them engage in productive discourse, and (c) attend a weekly pedagogy course focusing on practical techniques for enhancing learning. This pedagogy course is required for all first-time LAs. LAs that return for a second time (or more) become a part of the returning LA community, where they meet regularly and continue to discuss pedagogical strategies and best practices to engage students. LAs are paid approximately $15 USD per hour to work approximately 8 h per week working with students, preparing, and meeting with the lead instructor. They receive two credits for the pedagogy course. These values are slightly different at different universities. Although LA pedagogy courses differ throughout the world, they generally focus on at least one or more of the three units described briefly below (figure 11.2).

Each new LA enrolls in the LA Pedagogy Course concurrent with their first semester of being an LA. In this course, readings, assignments, class activities, and discussions are designed to help LAs implement active learning teaching practices. The main learning goals are: LAs contribute to a supportive learning community

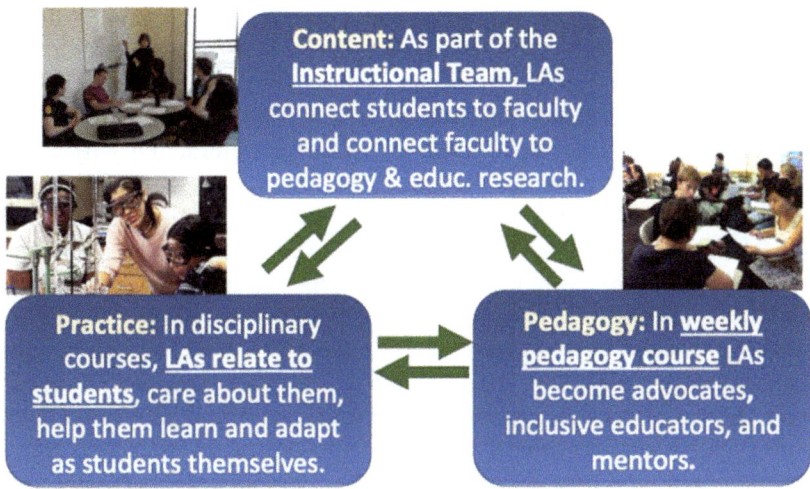

Figure 11.2. The three pillars of the LA model. LAs engage in content planning with the lead instructor of the course as a part of the instructional team, they engage in practice as they interact with students, and they take a practicum-based weekly pedagogy course that draws on their LA experiences.

and: (a) build knowledge about how they and their students learn, (b) practice listening and questioning, eliciting and responding to student ideas, and engaging students in peer interactions that support learning, (c) develop and practice strategies for facilitating equitable, inclusive, and effective learning environments, and (d) reflect on their learning, teaching, and views of effective education.

To support these goals, the course is divided into three broad units:
- **Unit 1—What to do:** Skills and strategies for supporting student learning include: questioning, wait time, facilitating student–student interactions, compassion, listening, and how to give effective feedback.
- **Unit 2—When to do it**: Learning theory helps LAs make instructional decisions regarding how, when, and why to use the skills learned in unit 1. This includes discussing, noticing, and eliciting student ideas, formative assessment, and learner-centered teaching.
- **Unit 3—Why it matters:** The purpose and impacts of education is explored near the end of the course, including discussions around belonging and inclusion, metacognition, the intended and unintended consequences of education, and impacts of the LA model in higher education.

LAs' placements in STEM courses serve as a practical, field-based component of the pedagogy course, where they collect relevant observations and analyze them through the literature on student learning. At the same time, the pedagogy course prepares students for their work in these STEM courses by helping them reflect on their interactions with students, setting goals, and evaluating their own effectiveness. There are three different practicum assignments. The first, called 'Getting to know your students,' takes place early in the semester and requires that LAs survey the students in the course to determine who they are, how they feel about the course, or any other questions that the LA feels will help them learn about the student population with which they are about to work. They then analyze their data and present their findings in graphs, tables, and narratives and use this to set goals for how they can best support students' learning. The second practicum assignment is an audio transcript where LAs audio record themselves interacting with a group of students in the class for which they are an LA. They then transcribe their audio, read it, get feedback from fellow LAs about their interaction, and ultimately annotate the transcript to describe ways in which they used skills introduced in the pedagogy course and things they wish they had said and done. They then write a reflection of this interaction and return to it later in the term and discuss how they have grown. LAs use this transcript as a concrete tool for establishing strategies for interacting with students and considering further improvements to their use of pedagogical techniques. This activity also promotes conversations about the learning environments and opportunities in different departments and encourages LAs to translate teaching best practices across disciplines. The third practicum assignment focuses on LAs receiving mid-term feedback from their students so that they can make changes and test new strategies. Again, LAs collect and analyze survey data and present it in a variety of forms. They discuss challenges and successes and make plans for improving for the remainder of the term. In the final practicum assignment. All new LAs are matched with an experienced LA who receives special mentoring training. These LA mentors meet weekly with their eight mentees and observe them at least once per semester as they work with students. Undergraduate LA mentors receive intensive training on mentoring and support from LA program staff. At the same time, they are the eyes and ears of the LA program and bring back to program personnel how LAs are actually being integrated into classrooms, faculty members and LAs who need more support, and the ways in which courses are organized throughout campus.

One of the things that makes the LA model different from other models that engage undergraduates in supporting their peers is that LAs work alongside instructors to prepare for and implement the course. Preparing undergraduates to serve on instructional teams requires an infrastructure that not only prepares LAs but also engages instructors in a variety of ways. Instructors seeking to use LAs often need a bit of time to try ideas, guidance in thinking through course designs and learning theories that support them, and information about multiple ways that courses can be designed for LAs to interact with students. Thus, when an LA program starts up on a campus, it supports instructors in course innovation and begins a process of departmental and institutional change. The LAs also carry important information into the awareness of course instructors, who are often researchers and do not typically have time to keep up with the literature on education.

The LA model was officially launched at the University of Colorado Boulder (CU Boulder) in 2001. Currently, approximately 450 LAs are hired each year in over 100 different courses in 18 departments within five colleges and schools. The program impacts roughly 25 000 student seats each year for a cost of approximately $55 per impacted seat (there is some duplicated head count). To receive LAs, faculty must submit a proposal where they articulate their thoughts about the purpose/process of instruction, the reason for proposed course changes, and their methods for evaluating effects of changes. The full proposal and LA application and hiring process is managed through special *LA Campus* software that was established by the team at CU Boulder but is used at universities throughout the United States. Since all instructors are welcome to apply to receive LAs at CU Boulder for an upcoming term, the LA model has spread beyond STEM and into humanities and social sciences. LAs work in introductory and upper division courses in physics, chemistry, biochemistry, biology, languages, business, history, English, engineering, psychology, media, communication and information, environmental science, writing, mathematics, astrophysics, and computer science. On some campuses, LAs are incorporated only in one department, although programs tend to expand to include other departments over time.

The use of the LA model has grown throughout the nation and throughout the world. In 2009, an International Learning Assistant Alliance (LAA) was formed to support hundreds of universities that use, or sought to use, LAs. To date, there are over 2000 individual accounts from over 500 institutions in 28 countries registered with the LA Alliance. There are many reasons the LA program has grown to this extent. One of these reasons is systematic research that demonstrates multiple facets of its effectiveness.

11.2 Research on the effectiveness of the LA model throughout the US

Research has demonstrated increased learning outcomes for students enrolled in LA-supported courses. Sellami *et al* (2017) studied LA- and non-LA-supported introductory molecular biology courses and through regression analysis found that students enrolled in LA-supported courses had higher average scores on common exam questions categorized in the upper tiers of Bloom's taxonomy. Introductory physics courses transformed with LAs show conceptual learning outcomes nearly twice the national average for traditional courses on similar assessments (Pollock 2009). Pollock also found longitudinal effects by measuring learning outcomes at the end of a junior-level electricity and magnetism (E&M) course taken by physics majors who did and did not have an LA-supported introductory E&M course as first-year students Those from LA-supported courses outperformed their peers who did not as freshmen have an LA-supported course. Those who had served as LAs in earlier terms outperformed everyone in the upper division class, and while selection effect was noted, students in the upper division physics course showed many academic similarities. In applied mathematics, Nelson (2011) demonstrated that students who were labeled 'at risk' by virtue of their scores on a mathematics entry exam were shown to be indistinguishable from their peers on a similar exam after participating in an LA-supported calculus course. Margoniner *et al* (2020) demonstrated that students in LA-supported, flipped courses outperformed their peers who were in traditional lecture courses.

Van Dusen and Nissen (2020) examined institutional data from 2312 students attending a Hispanic Serving Institution. Using hierarchical linear modeling and controlling for several variables including instructor effect, they found significantly lower failure rates for courses that used LAs than for those that did not. They also found that these failure rates were most decreased for student groups traditionally underrepresented in science. In another study that used data from the Learning About STEM Student Outcomes database from multiple universities, Van Dusen and Nissen (2017) found that while inequities between different groups exist, LA-supported courses are associated with greater learning gains for all students, and the relative differences in growth between groups are decreased through the use of LAs. Sellami et al (2017) demonstrated better learning outcomes on traditional exam questions in courses with LA support and this growth was shown to be even greater for students from traditionally underrepresented groups.

Alzen et al (2018) demonstrated that exposure to LAs in a course can impact retention in gateway courses generally. They used institutional data from 32 071 students at CU Boulder and demonstrated that being in at least one LA-supported course decreased students' failure rates by up to 60%. This study used institutional data from 16 cohorts of students who entered the university as full-time freshmen each fall from 2001 to 2016 and took Physics I/II, General Chemistry I/II, or Calculus I/II. Students were compared on the basis of race/ethnicity, gender, first-generation status, financial aid, number of credits upon enrollment, high school grade point average (GPA), and admissions test scores. This study also controlled for instructor effect.

Studies have also found that the LA program improves and enriches students' experiences in the classroom. A mixed-methods study that analyzed audio transcriptions of student in-class discussions using global and line-by-line coding revealed that students tend to engage in more discussion and ask better questions when an LA is present, and LAs who use quality questioning strategies are able to engage students in deeper thinking (Knight et al 2015).

Evidence suggests that the LA program leads to growth among LAs themselves. Close et al (2016) analyzed LAs' written responses and interview responses. They found that LAs engage in multiple communities within an institution in ways that help LAs become more integrated in disciplinary communities, leading to increased identification with science. Using institutional data, Otero (2015) found that of the 173 LAs included in their study, 168 (97.1%) graduated within six years, compared to 88.3% ($n = 9215$) of students in the sample that was matched to the LAs by gender, predicted GPA, class level, and college enrollment. Margoniner et al (2020) also found that LAs reported that their experience as an LA both improved their understanding of learning and helped them become better learners.

In teacher recruitment and preparation, it has been demonstrated that the number of science and math majors completing the teacher certification program nearly tripled by comparing 4-year averages before and after the LA program began recruiting LAs into teacher preparation programs (Otero 2015). A longitudinal study of LAs who became K-12 teachers ($N=15$) was conducted from 2006 to 2011, comparing teaching practices of former LAs to those of a matched sample of teachers who went through the same teacher preparation program, had similar GPAs, taught in the same schools, had similar majors, but did not serve as LAs as undergraduates (*N*=14). In most areas, LAs showed significantly higher use of research-based instructional strategies than the matched sample of teachers (Gray *et al* 2016). Regarding teacher recruitment, (Margoniner *et al* 2020) found that 73% of their 58 LAs reported that they became interested in teaching as a career after participating as an LA.

The literature regarding LAs continues to grow measuring a variety of things from impacts on students to effects on faculty. Studies have also shown that combined with other research-based practices, the LA program shows promise in promoting institutional change (Goertzen *et al* 2011), including changes in faculty members' roles and identities and large-scale changes in funding allocations and general educational practices and physical classroom settings. In a qualitative study, McHenry *et al* (2009) found that faculty and LAs grew in a variety of categories including: expanded conceptions, confidence as teacher and learner, collaborative benefits, course design and teaching process, relationships and interdisciplinary connections. This study reveals ways in which interacting with an LA program leads to potentially sustainable change for faculty members.

Studies about how LAs interact with students are also emerging in the literature. A recent study by Thompson *et al* (2020), documented how 25 LAs interacted with their students in introductory level and upper division biology, chemistry, and physics courses. LAs wore video cameras for an entire class day and the recorded interactions were used to inductively develop the Action Taxonomy for Learning Assistants (ATLAs) which categorized LA classroom moves. They found that LAs directed or guided classroom facilitation, gave advice or feedback, and engaged in course or non-course related talk, among other things (Thompson *et al* 2020).

11.3 Research that suggests reasons for why the LA model is effective

Researchers have begun to ask how the LA model can produce outcomes such as those discussed above in multiple, diverse university contexts. In one study, Clements *et al* (2022) used a 'belonging' survey in LA-supported and non-LA-supported biology courses and demonstrated that students in the LA-supported courses revealed a greater sense of belonging than those without LAs and this finding was consistent in in-person and remote classrooms. Hernandez *et al* (2021) used the notion of 'social support' to create an explanatory model for LA effectiveness. By drawing on the literature, they first argue that social support has been shown to be associated with class engagement, social responsibility, school identification, valuing of learning, mental health, persistence, and identity development. Hypothesizing that LAs can provide the four types of social support (appraisal, emotional, informational, and instrumental), they established an instrument for measuring social supports in active learning. The instrument is called the Perceptions for Social Support for Active Learning Instrument (PSSALI). After their exploratory and confirmatory factor analysis they ended up with a 3-factor instrument that measures appraisal, emotional, and informational support. They found that students in an introductory chemistry class perceived strong social support from LAs. They then ran a regression analysis with findings from the FABUS classroom engagement survey (Brazeal *et al* 2019). They found a positive correlation between social supports and deep engagement in active learning in the LA-supported chemistry course.

Finally, Top *et al* (2018) demonstrated that students in LA-supported courses, and the LAs themselves, indicate that LAs provide a bridge between the student body and institutional structures, leading to strong connections to the discipline, to the university, and to the learning process and a sense of belonging. Top hypothesized that these connections were made possible because of LAs' dual identities as students and instructors. Students report that their undergraduate LAs are 'just like me,' and 'are going through the same thing.' At the same time, students and faculty view LAs as legitimate members of the instructional team. Thus, LAs provide the connective tissue for students—helping them feel like a part of, rather than excluded from, the class, the discipline, and the institution. This leads to dramatic impacts on student success, as reported above. In contrast, Top's study showed that students spoke of graduate student Teaching Assistants (TAs) as 'older,' 'in charge,' 'dictates the class,' 'intimidating,' and 'didn't really get what I was asking.' LAs' dual identities as students and as teachers creates helps to broker belonging and inclusion for students.

A national research agenda has emerged in the US and has led to multiple dissertations, an annual conference focusing on research on all aspects of the LA model. A recent article (Barrasso and Spilios2021) provides a scoping review of over 60 research articles that focus on many facets of the LA model. The International Learning Assistant Alliance website (LAA 2022) provides an up-to-date list of publications regarding the many facets of the LA model.

11.4 LAs are used in different contexts

Based on Faculty Course Proposals (FCPs) and information provided by LAs and their Mentors, LA-Supported courses at CU Boulder were categorized according to settings in which LAs interact with students. These categories are: (1) lecture, (2) recitation (additional group work session, usually required), (3) lecture + recitation, (4) laboratory, and (5) online/asynchronous settings. In many cases these experiences are supplemented with out-of-class, LA-facilitated office hours, study sessions, and help room sessions.

In lecture courses LAs support student learning by facilitating active learning in clicker questions and peer instruction (Knight *et al* 2015) group activities and discussions (Pollock 2009) and fully flipped classroom-style (Sellami *et al* 2017, Margoniner *et al* 2020). In one design, LAs attend all lecture course sessions (typically 3×/week) and facilitate group activity during some of the sessions (typically 15%–80% of class time during clicker questions and peer instruction). In another design, LAs attend one of two or three class sessions per week and facilitate student engagement 100% of those sessions, which are explicitly designed for group work and use of LAs, such as in a flipped classroom situation. In these situations, typically each LA supports 3–4 sections of the course, interacting with the same students each week in interactive group projects.

> LA support is essential to make the small group interactive learning activities (clicker questions and tutorials) effective for the students in a class of this expected size. I am also interested in exploring further into 'flipping the classroom', which may include giving students time in lecture to work on homework problems in groups, during which the LAs could help to answer student questions.
> (PHYS 2210 faculty member)

In another context, LAs interact with students in sessions that run in addition to the lecture. These are typically 1 h sessions for 5–30 students per session that run many times throughout the week. They are referred to as recitation sessions, group study hours, workgroups, and co-seminars. During these sessions, students work in groups of 4–5 on various problems or projects. LAs meet with the same groups each week and facilitate multiple recitation sections, optimizing the LA to student ratio.

> LAs are a vital component in our recitation sections as they aid us in the implementation of active learning projects through assistance in facilitating group projects. Our LAs are a great benefit for encouraging students to ask questions and discuss ideas in class since students are less intimidated by the LAs, in comparison to either the TA or instructor. The LAs also provide us a student's perspective when we discuss ideas that students might struggle with or ways to effectively implement activities.
> (MATH 1300 faculty member)

In another context, LAs interact with students during lecture and recitation/group work sessions, in relatively equal proportions. Typically, both of these course elements are required for students enrolled. LAs are also used in laboratories. These may be computer programming or research software labs, science labs, writing courses, or engineering design contexts. In these contexts students work in groups of 3–5 on hands-on experimental or studio projects and labs. Laboratory attendance and active participation is typically required of all students. Similar to recitations, LAs typically work with instructors and/or TAs to cultivate inclusive classroom environments. Programs have found that when the settings in which LAs interact with students are required, it is a better experience for students and LAs and more students benefit. Note that each of the sessions described above may take place in person or in online-synchronous settings. In synchronous sessions, LAs typically move between breakout rooms and engage students in discussions.

> I began using LAs in this course over 5 years ago and it was the best teaching decision I have made. Because of their contributions, we can hold many more Student Hours a week, provide more individualized instructional attention in lab, and provide students with an additional support system in a class that feels intimidating to many students. The 4000-level methods course requirement is designed to be one of the more (most) challenging courses a student takes in our major so the more ways we can help students succeed, the better. Specifically, LAs work with students on lengthy and challenging Article Assignments, weekly quizzes, lab sample problems, exam review, and the final lab project. Students also have indicated that they like it when LAs help teach some of the lab materials.
>
> (PSYC 4443 faculty member)

Finally, LAs interact with students in online, asynchronous courses. In these settings, they interact almost entirely asynchronously via discussion boards, PDF-annotation tools like Perusall, chat-style tools like GroupMe, e-mail, videos, etc. In these courses, students are required to engage in conversation with the LA and with one another, as an essential source of feedback during the learning process.

There are additional LA roles. Most LA appointments include responsibilities in addition to working with students during class time, such as help rooms, LA-led study sessions, and office hours. In these settings LAs support students one-on-one and/or in small groups. Help rooms are student support resources at a specified location where students can find help during most weekday hours. LAs often spend 1–2 h per week helping to staff help rooms. Notable examples include the Math Academic Resource Center (MARC) and Physics Help Room, both are highly utilized and appreciated by LAs and students alike and members of the Academic Resource Center Network. LA-led study sessions are typically held when an LA feels that it is important or in coordination with their lead faculty around exam review schedules. These range from formal to informal sessions with attendance ranging from 10 to hundreds of students. LAs typically consult with their lead instructor or pedagogy course instructors. Office Hours may take place in a departmentally-specified location, informal spaces on campus, off campus, or online. These hours are not generally publicized beyond a particular course section, and students typically only visit the LA specific to their course section. Examples are provided below:

- **'Teach an LA':** These activities were pioneered in the psychology department, and have since spread to ENGL and HIST. Students are required to visit an LA for 10 min once per unit to talk through their understanding of a piece of course content. This structure encourages students to build relationships with their LA(s) and one another.
- **LA check-ins:** Students are expected to sign up for one or more of these 10–15 min meetings within the first few weeks of the semester to meet the LA and ask

questions about course structures and content. These meetings are a great way to build connection and community, particularly in asynchronous courses.
- **Homework:** In courses with challenging homework assignments, students tend to utilize LAs more frequently. This is particularly true in computer science where novice programmers struggle with learning how to debug their code.
- **Quiz-checks:** Students are able to meet with an LA to talk through their responses to a content quiz and earn back some points, with goals of helping to develop student metacognition.

11.5 A model of institutional change

Since its inception in 2000, the LA model has been a model of institutional change (Otero 2015). The designers of the model have considered it a mission for faculty learning and for the evolution of instructional environments. The LA model is an iterative model of change, meaning that it respects classic models and practices, supporting their co-existence, while iteratively transforming department-level, and campus-level practices and structures. Through this process, stakeholders transform values and expectations for what it means to educate and to be educated (Sewell 1992). At the same time, the LA model opens new possibilities for faculty both through their own experimentation and through their interactions with other faculty and the LA program. Through the LA community they have access to resources, including LAs, allowing them to consider what could be rather than 'what is' or what 'has been' in instruction.

> Being an LA means being the bridge between the knowledge of the Professors and the student experience. Even the best teachers can't connect with every student, and it can sometimes be much easier to bond with an educator who is at the same level as you.
> (LA)
> Fellow undergraduates are less intimidating, and potentially have an easier time forming relationships with students which encourages excitement and curiosity.
> (LA)
> I think our main job is to help students feel more comfortable asking questions and learning in big classes where sometimes it is easy to feel alone and voiceless.
> (LA)
> Being an LA means being a mentor for people of a similar age to reach out to without feeling afraid of the fear of being embarrassed.
> (LA)
> The role of an LA is also to help create an environment conducive to learning, meaning that it is our responsibilities as LAs to create environments free from judgment, discrimination, and harassment

An important difference between the LA model and other course transformation efforts is that the LA model does not promote one particular idea of what a healthy course environment looks like. Rather, following complex adaptive systems models of change, it understands that the community seeks to bring together an interrelated set of diverse individuals to work on problems that have yet to be defined.

11.6 Get involved: the International LA Alliance

LAs, LA Mentors, and faculty also have opportunities to engage in wide-spread curricular development and transformation through participation in the LA Alliance via regional and international workshops and conferences. These workshops are collaboratively developed and facilitated by LA stakeholders from across the country and around the world and create spaces for collaborative disciplinary and interdisciplinary discussions about how to utilize LAs as a tool for promoting collaborative, active engagement in all classrooms.

11.7 LA Alliance: https://learningassistantalliance.org/

11.7.1 LA program at CU Boulder: https://www.colorado.edu/program/learningassistant/

Through Regional Workshops, onsite and virtual campus workshops, an annual international conference and LA research symposium, and multiple special-interest LA Slack channels, LA users collaborate, share, and learn from one another. The vision of the LA model is: to efficiently build lasting capacity among faculty, courses, and departments for sustained offerings of high quality, learner-centered instruction. In these settings every student feels included and valued and is comfortable accessing multiple forms of support in and outside of the classroom. Undergraduate students who serve as LAs and LA Mentors become effective leaders, teachers, and team members, prepared for the increasingly diverse and interdisciplinary workforce.

The mission of the LA Model is: to provide infrastructure (human and technological) necessary for improving student success by increasing the diversity of university course instructional teams through the inclusion of pedagogically trained, undergraduate Learning Assistants (LAs). Ongoing development opportunities and communities are available for faculty, departments, and undergraduates leading to growth and development as educational leaders, mentors, and state of the art educational innovators. The LA model is an efficient, effective, and equitable program for building and maintaining excellent experiences for students and faculty across university campuses.

11.8 Final thoughts

Through the semester-long pedagogy course, their interactions with students, and through their weekly meetings with the lead instructor of the course, LAs carry important messages throughout campus. The LA Program also carries important messages throughout campus both through frequent interactions with faculty and through LA Campus program management software. Themes from these messages are highlighted below.

Inclusive classrooms: LAs are a part of the instructional team. They provide direct support to faculty and students. They relate to students, give them voice, care about them, and help them learn. They meet weekly with the lead instructor to provide insight on how students are experiencing the course and to plan for the upcoming week.

> Working with LAs has helped me refresh my teaching strategies and resist the temptation to just do what's worked in the past. I enjoy helping the LAs take on more responsibility and gain confidence in their leadership skills, and in turn, this experience reminds me of the greater purpose and goals of education.
>
> (English Faculty member)

Faculty development and classroom innovation: Faculty who work with LAs participate in workshops and receive weekly newsletters that focus on pedagogy, active learning pedagogies, course design, use of LAs, and inclusivity. In the online Faculty Course Proposal (FCP) faculty articulate course design, innovation ideas, plans for using LAs, and plans for assessing outcomes. Through the LA Campus software (developed at CU Boulder), faculty then receive feedback first from their Departmental Coordinator (DC) and then from the campus LA Program Coordinator (PC), who meets regularly with the DCs.

System of innovation: Faculty reflect on their innovations throughout and after implementation. They reflect individually, with LAs, with other faculty and LA Program staff. Their learning is reflected in their subsequent Faculty Course Proposals.

> My work as an LA involves working with the university to provide meaningful feedback on their method of instruction, their students and the effectiveness of the LA program as a whole
>
> (LA)

Individualized faculty mentoring by LA Program staff: The LA Program builds relationships with hundreds of faculty members throughout campus and continually builds knowledge of instructional innovation across hundreds of CU Boulder courses. Thus, LA Program staff members are able to provide timely, relevant e-mail, Zoom, and in-person consulting for faculty including course design, use of LAs, best practices with Canvas, remote instruction, and more.

LAs are supported by a weekly pedagogy course or Returning LA community and by undergraduate LA Mentors, who observe and consult with LAs. The pedagogy course introduces skills such as questioning, listening, wait time, feedback, compassion, inclusion, metacognition, and learner-centered teaching. Faculty receive a weekly newsletter with the pedagogy concept of the week, suggestions for their meetings with LAs, and other program highlights and events. Undergraduate LA Mentors receive intensive training on conferring, consulting, and mentoring. They meet weekly with their 'PODs' of LAs and observe them and provide feedback.

Undergraduate LA Mentors are the eyes and ears of the LA Program: They provide insight regarding classroom implementation, providing early alerts of areas of interest and concern. This often leads to faculty consultations, facilitated by LA Program staff regarding needed improvements in course designs, uses of LAs, and improved student experiences. LA Mentors also raise awareness of successful innovations, allowing LA Program staff to spread good ideas and provide recognition for faculty and departmental successes.

> Watching my LAs progress from the beginning of the semester where they're unsure how to even be an LA, to the end, when they look like seasoned pros, and I think that's really one of the most rewarding parts, knowing that I had even a little bit of a hand in doing that.
>
> (LA Mentor)
>
> I've become more aware of the experiences of students outside of my department. It gives me a better impression of what the university experience is like as a whole for a lot of people. The most interaction I've had outside of my department is in the LA Program, so it gives me a more holistic view of the school and the programs that are out there.
>
> (LA Mentor)

The LA infrastructure provides an efficient mechanism for continued growth of faculty and course designs, and dissemination of educational innovations throughout campus. Hundreds of faculty members engage each year and share innovations with one another. This leads to a sustained capacity for improved student success for a great diversity of students at CU Boulder.

> As it's my last semester being an LA, I just want to express its immeasurable what this program has done for me. I think the community, the pedagogy, the mission—it's all made me a better person. I obviously could use my LA experience on resumes and stuff, but I think it extends way further than that. Thank you so much for everything.
>
> (Returning LA)

References

Alzen J, Langdon L and Otero V 2017 The learning assistant model and DWF rates in introductory physics *Physics Education Research Conf. Proc.* (Melville, NY: AIP Press)

Alzen J, Langdon L and Otero V 2018 A logistic regression investigation of the relationship between the learning assistant model and failure rates in introductory STEM courses *Int. J. STEM Educ.* **5** 56

Barrasso A and Spilios K 2021 A scoping review of the literature assessing the impact of the LA model *Int. J. STEM Educ.* **8** 12

Brazeal K R, Brown T L, Brassil C and Couch B A 2019 Cultivating active learners: how instructors can change their teaching to help students engage with formative assessments *Society for the Advancement of Biology Education Research National Conf.*

Clements T, Friedman K, Johnson H, Meier C, Watkins J, Brockman A and Brame C 2022 'It made me feel like a bigger part of the STEM community': incorporation of learning assistants enhances students' sense of belonging in a large introductory biology course *CBE—Life Sci. Educ.* **21** 1–13

Close E W, Conn J and Close H G 2016 Becoming physics people: development of integrated physics identity through the learning assistant experience *Phys. Rev. Phys. Educ. Res.* **12** 010109

Crouch C H and Mazur E 2001 Peer instruction: ten years of experience and results *Am. J. Phys.* **69** 970–7

Goertzen R M, Brewe E, Kramer L H, Wells L and Jones D 2011 Moving toward change: institutionalizing reform through implementation of the learning assistant model and open source tutorials *Phys. Rev. Spec. Top.-Phys. Educ. Res.* **7** 020105

Gray K E 2013 Teaching to learn: analyzing the experiences of first-time physics learning assistants *Doctoral Dissertation* (University of Colorado at Boulder, Boulder, CO, USA)

Gray K E, Webb D C and Otero V K 2016 Effects of the learning assistant model on teacher practice *Phys. Rev. Phys. Educ. Res.* **12** 020126

Hernandez D, Jacomino G, Swamy U, Donis K and Eddy S L 2021 Measuring supports from learning assistants that promote engagement in active learning: evaluating a novel social support instrument **8** 2–17

Knight J K, Wise S B, Rentsch J and Furtak E M 2015 Cues matter: learning assistants influence introductory biology student interactions during clicker-question discussions *CBE—Life Sci. Educ.* **14** 1–14

LAA. 2022 *International Learning Assistant Alliance* https://learningassistantalliance.org

Låg T and Sæle R G 2019 Does the flipped classroom improve student learning and satisfaction: a systematic review and meta-analysis *Am. Educ. Res. J. Open Access* **5** 1–17

McHenry N, Martin A, Castaldo A and Ziegenfuss D 2009 Learning assistants program: faculty development for conceptual change *Int. J. Teach. Learn. High. Educ.* **22** 258–68

Margoniner V, Bürki J and Block M 2020 Learning-assistant-supported active learning in a large classroom *Am. J. Phys.* **88** 923–33

Nelson M 2011 Oral assessments: improving retention, grades, and understanding *PRIMUS* **21** 47–61

Otero V 2015 Nationally scaled model for leveraging course transformation with physics teacher preparation: the Colorado learning assistant model *Effective Practices in Preservice Teacher Education* ed E Brewe and C Sandifer (College Park, MD: American Physical Society and American Association of Physics Teachers) pp 107–16

Pollock S 2009 Longitudinal study of student conceptual understanding in electricity and magnetism *Phys. Rev.: Phys. Educ. Res.* **5** 1–8

Sellami N, Shaked S, Laski F A, Eagan K M and Sanders E R 2017 Implementation of a learning assistant program improves student performance on higher-order assessments *CBE—Life Sci. Educ.* **16** ar62

Sewell W H 1992 A theory of structure: duality, agency, and transformation *Am. J. Sociol.* **98** 1–29 https://www.jstor.org/stable/2781191

Seymour E, Hunter A B, Thiry H, Weston T J, Harper R P, Holland D G, Koch A K and Drake B M 2019 *Talking about Leaving Revisited: Persistence, Relocation, and Loss in Undergraduate STEM Education* (Berlin: Springer)

Thompson A N, Talbot R M, Doughty L, Huvard H, Le P, Hartley L and Boyer J 2020 Development and application of the action taxonomy for learning assistants (ATLAs) *Int. J. STEM Educ.* **7** 1

Top L M, Schoonraad S A and Otero V K 2018 Development of pedagogical knowledge among learning assistants *Int. J. STEM Educ.* **5** 1

Van Dusen B and Nissen J 2017 Systemic inequities in introductory physics courses: the impacts of learning assistants *Physics Education Research Conf. Proc.* (Melville, NY: AIP Press)

Van Dusen B and Nissen J 2017 Systemic inequities in introductory physics courses: the impacts of learning assistants *Paper Presented at Physics Education Research Conf. 2017 (July 26–27, 2017) (Cincinnati, OH)*

Van Dusen B and Nissen J 2020 Associations between learning assistants, passing introductory physics, and equity: a quantitative critical race theory investigation *Phys. Rev. Phys. Educ. Research* **16** 010117

Van Dusen B, White J and Roualdes E 2016 The impact of learning assistants on inequities in physics student outcomes *Paper Presented at Physics Education Research Conf. 2016 (July 20–21 2016) (Sacramento, CA)*

IOP Publishing

Effective Teaching in Large STEM Classes

Anna K Wood

Chapter 12

Effective teaching in large classes; looking through and beyond the COVID-19 pandemic

Simon Bates and Firas Moosvi

The final chapter of this volume examines how the COVID-19 pandemic dramatically re-shaped our large class instruction, catalysing innovations already in play and simultaneously requiring a rethink of the way we approached the design and delivery of courses. We focus on the areas of course content and its delivery, class community and a sense of belonging, the way students and instructors were able to interact with each other, and the many challenges in assessment of learning. As we emerge from the most severe disruptions brought by COVID-19, we offer some insights as to how these fundamental ingredients of a course may evolve in the future.

12.1 Introduction

This chapter applies the lens of 'experiences gained from the COVID-19 pandemic' to offer some thoughts around where we are now with respect to effective teaching in large classes, and where we may be going as a result. The disruptions brought by COVID-19 have the potential to radically alter teaching and learning, encompassing areas including (but not limited to) dimensions of learner expectations, instructor capabilities, and institutional models for teaching and learning. Whilst the experience of responding to the various and complex challenges of COVID-19 has certainly challenged the status quo, it has also had a catalytic effect on things already in progress, and driven them rapidly towards much wider scale adoption.

Much by way of enhancement and innovation was certainly underway in this domain pre-COVID-19, but the successive and sustained challenges to our traditional teaching and learning model during COVID-19 accelerated a great many aspects, as well as requiring us to consider completely new scenarios. We focus in this chapter on the following sections: we first consider the various dimensions of the post-COVID-19 context which we find ourselves moving towards, before turning to the fundamental building blocks of a course, in terms of operations and logistics; curation and presentation of content; interactions with content, peers and instructors; and assessment.

12.2 The emerging post-COVID-19 context

The challenges to teaching and learning brought by COVID-19 were multiple waves of uncertainty and disruption, requiring a series of large-scale changes to the way student learning took place at our institutions. The initial 'pivot' to fully online in March 2020 was rapid and chaotic, as academic staff and students grappled with upheaval in their personal and academic lives: we did not do online learning, we managed as best we could to deliver emergency remote instruction (Hodges *et al* 2020). There was much to learn for many academic staff, and effectively zero time to do so, as the pivot occurred midway through an academic term. A particular challenge came from the way curricula are structured in the physical sciences; topics are presented linearly and learning is linear and additive, in the sense that concepts and content build on what has been covered before, meaning that the choice to excise content from the remaining weeks of the course was not a particularly viable option. Subsequent courses would require at the very least basic familiarity with content from courses taught earlier in a sequence.

This pivot to emergency remote instruction was followed by an intense period of redesigning for fully-online instruction, when it became clear that the disruption from COVID-19 was going to extend well into the 2020–1 academic year. Some programs, such as those in the health and clinical sciences continued in hybrid mode, with logistical challenges such as physical distancing requirements and the need for personal protective equipment in classrooms and clinical settings. For most instruction though, this was fundamentally a course redesign process conducted at unprecedented scale; essentially every member of academic staff (re-)designed and delivered courses in alternate modalities to the predominant face-to-face model pre-COVID-19. This was accompanied by a greatly amplified range of discussions around teaching and learning challenges and practice, and—for our institution—strong institutional support in terms of additional resources to deploy within programs and departments to best fit local needs. This combination of circumstances and actions represented a fundamental phase change in how we thought about, talked about and delivered our taught courses. The challenge, moving forwards, is to chart a deliberate path towards a 'new normal' that retains some of the positive aspects learned through this period, whilst at the same time mitigating the challenges.

One particular aspect that was brought into sharp focus during this period was inclusion—a sense of belonging and connection to the institution and academic programs. It has been reported (Napierala *et al* 2022) that many learners struggled with the personal and social isolation from learning extensively online, and that these challenges were very unevenly distributed across different groups of learners. In contrast, it was also the case that some learners—particularly those with academic accommodations on accessibility or disability grounds—felt more productive and more engaged with their peers, their instructors and more part of an academic community. Preparedness of new students entering Higher Education, following a period towards the end of high school of essentially fully online learning, was a concern, though one study reports essentially unchanged distributions of scores on an introductory diagnostic test to determine course placement, comparing pre- and during-COVID-19 cohorts (Burkholder and Wieman 2022).

12.3 Course design and logistics

As other chapters in this volume have outlined, there has, over the last two decades, been a significant shift in course design in terms of re-thinking the ways that classroom time, particularly lecture time in large introductory courses in Physics and other STEM subjects, has been used. Active learning pedagogies have been widely adopted across many disciplines, with significant impacts on learning and student retention (Freeman *et al* 2014) (see chapter 2 for more detail). Pedagogies utilizing a variety of approaches are discussed in chapters 3 and 7–10. These include, but are by no means limited to, the use of clickers, worksheets for small group problem solving activities, classroom discussion and demonstrations. Such approaches are now well-established instructional designs across most institutions, with a convincing corpus of research evidence and local practice and experience. This use of in-class time has also been augmented by increased use of instructional media and textbook pre-readings in a flipped-classroom approach, where students develop basic content familiarity ahead of coming to class, and class time focuses on building conceptual understanding and mastery rather than didactic presentation of content.

Similarly, additional staff resources were used to support course logistics and the additional demands on instructors. This is not dissimilar to the team-teaching model, increasingly common in large courses pre-COVID-19. Often graduate teaching assistants (or, increasingly, undergraduate teaching assistants) were deployed to support the delivery of courses, with a range of duties including marking, lab and tutorial supervision, but also as 'lecture TAs' (supporting the instructor in interactive engagement activities in active learning classrooms). For more detail on the use of undergraduate TAs through the learning assistance model, see chapter 11. The advent of COVID-19 expanded both the range of support to include learning technology roles, instructional design and student support/community engagement activities, to name but a few. Teaching these large courses became solidly established as a team-sport, with various skills and competencies required of the teaching team as a whole (Bates 2014).

Learning technology tools were integral to supporting these re-designs, from the obvious wider uptake of the institutional virtual learning environments (VLEs), through to ways to include geographically dispersed learners (e.g., the use of a cloud-hosted version of electronic voting systems, permitting 'from-anywhere' access, rather than handsets tied to physical base stations in classrooms). A variety of 'just-in-time' tools and applications permitted the opportunity for instructors to gather feedback, input comments from students, both synchronously in class meetings and before/after scheduled classes as well.

For all potential benefits that this redesign brought, there were also many challenges to navigate. A distributed (in some cases, global) cohort of learners presented logistical challenges that were beyond the direct control of instructors. These included time zone differences for synchronous classes; access to stable networks with sufficient bandwidth and access to quiet and private spaces in which to participate in class sessions or summative assessments. Navigating (and mitigating) these challenges fell first to the instructors teaching these courses; class recordings, flexible submission deadlines and course concessions were widely used to support students in these circumstances. Small group discussions, a cornerstone of active learning pedagogies in lectures, workshops/studio style tutorials were clunky and inefficient in breakout rooms, with the technology clearly impeding the natural flow of collaboration, conversation, and discussion. Trying to collaboratively solve problems on a shared whiteboard in Zoom or Teams felt significantly less fluent than gathering around a physical writing surface, such as a piece of paper or a whiteboard. Many academic staff paid the price at the time to do what was necessary to ensure the best possible continuity of learning for their students, only counting the cost of doing so as the pandemic disruptions stretched into months, then years. This has taken a toll, with faculty burnout and a detriment to their own health and well-being being common.

12.4 Course content

The disruptions from COVID-19 made it abundantly clear—were it ever in any doubt—that an instructor's role in a course is far more than delivery of content. That said, content creation and curation was one of the easier components of a course to re-factor for a different modality; presentation to a class was replaced by presentation to a webcam, either live to cohort or by way of pre-recorded digital content in smaller-than-a-whole-lecture chunks of time to be reviewed (and, often, re-reviewed) by learners. Creation and use of open content (Open Educational Resources) likewise increased, as people sought out efficiencies, shared materials, and tried not to reinvent wheels. Much of this material remains available for re-use post-pandemic, encouraging a sustainment of versions of blended, hybrid, and flipped learning approaches. An instructional choice that required some deliberate thought from the instructor was to achieve the right balance of flexibility versus community for course content, balancing synchronous (whole class time) versus asynchronous (in your own time) activities within a particular course. Recording lectures, which offers flexibility plus the opportunity to rewatch and review at a later date, can also lead to student disengagement and poor learning practices, such as not keeping up with material but instead cramming immediately prior to an exam. There are affordances and limitations to recording lectures, and it remains an instructor and disciplinary choice whether this fits with the learning design of a particular course (O'Callaghan *et al* 2017). Some elements of active learning that support deep engagement with course content were able to be translated to online and asynchronous settings. Cloud-hosted electronic voting systems supported interactive engagement approaches. The real-time peer-discussion components were arguably more challenging in virtual spaces, but asynchronous approaches to peer discussion and response were also successfully deployed (Bruff 2021).

12.5 Interactions

Interactions and discussions, whether between students as peers in a cohort or between students and the teaching team, are a foundational aspect of an effective and engaging learning environment. Pre-COVID-19, it was all too easy to take the humanity of teaching for granted, along with the many forms of in person interaction that took place, from scheduled formal interactions to informal and serendipitous exchanges. During COVID-19, we saw just how much students needed these opportunities. Though academic staff worked hard to create opportunities for discussion and interaction in online spaces, the absence of the more spontaneous connections and interactions often led to disengagement from a learning cohort.

There were several unexpected positives as well: the chat backchannel in common conferencing platforms like Zoom and MS Teams enabled a 'lower barrier' method to communications, supporting engagement from students who would otherwise may not want to raise their hand in class. This fostered not only engagement but also inclusive participation. In large classes, technology tools meant that 'everyone could be in the front row' (Mazur 2022). The other side of this coin was that in large classes, the volume—and sometimes, relevance—of a backchannel stream needed dedicated effort to manage and curate, often a key role for a lecture TA in large classes. Keeping discussions relevant, and on topic, could be a challenge as the formal instructional tools blurred with personal and social communications: everything was just 'online'. However, these were certainly fertile grounds for discussion relevant to course content, as well as building community and peer support in difficult circumstances, which grew healthily with appropriate facilitation. As had been the case long before COVID-19, students used personal and social platforms (e.g., Discord) to connect and supplement interactions in institutionally-provisioned spaces. A further challenge was that different courses often used different platforms and applications for interactions within a given course, leaving students often asking why we couldn't all just agree on a single tool or application. Choice is a good thing; too much choice might not necessarily be so, particularly when it feels overwhelming when looking across a sequence of courses taken by an individual student simultaneously.

Learning analytics—the use of data captured and analyzed during student use of learning applications to support actionable insights—offers significant potential here, to be able to understand and enhance learning activities and interactions at both the individual (personalized) and cohort level. There's also the potential for efficient communication if deployed appropriately in large classes. Here, the use of multiple tools and applications within a course to support student learning leads to this data being fragmented across different systems. Without an automated process at scale to be able to merge these different data sources—and at the same time ensure the integrity of the merged data—the insights and actions that follow from them have tended to be limited to those analyses and insights located within a single learning system, most often the VLE. Curating the data manually to be able to take action was often simply not feasible with all the other demands on an instructor's time.

Most major VLEs offer at least rudimentary analytics and some actions that can be taken as a result, for example a low-barrier but somewhat limited impact approach, is 'nudge' emails to students who did not submit particular assignments, or whose performance on assignments may give cause for concern. More sophisticated and flexible systems exist such as OnTask, a learning analytics tool that 'empowers educators to collect, collate, analyze, and use student engagement and success data that they consider meaningful for their particular contexts ... [and] enables personalization and targeting of student learning and support ..., fostering positive student–teacher relationships and enhancing student engagement' (Pardo *et al* 2019). Such interventions have been shown to act as beneficial 'nudges' to change student behaviours, albeit one with a time overhead to set up the environment, define the rules and regularly import learning data. Both Paredo *et al* and Vigentini *et al* show the positive impact of such learning analytics tools on student motivation and academic achievement (Vigentini *et al* 2017, Pardo *et al* 2019).

12.6 Assessments

Two other chapters in this volume take a deeper dive into the nature and format of assessments: chapter 5 examines authentic and inclusive assessments, and chapter 6 discusses the use of computer marked assessments and concept inventories. Assessment of students in large classes was arguably the most challenging area to navigate during the pivot to large-scale online teaching and learning during COVID-19. Whilst online assessments in large classes were prevalent pre-COVID-19, they tended to be used for weekly homework problem sets, practice problems and reading comprehension quizzes ahead of discussion of material in subsequent in-person classes. Far less common was the use of online assessments for high-stakes summative assessments such as midterms or final examinations: the unseen, time-limited, on-paper examination was by far the predominant assessment model.

The challenges of COVID-19 thus required a complete rethink. A number of different approaches were devised and deployed, such as 24 h take-home exams, project/portfolio-based assessments, purely online quizzes, and assessments which required students to upload written answers and working via document-capture through phone apps or cameras. Class size was a significant factor in determining which assessment methodologies were feasible in resource terms, often tensioned against pedagogical goals. What might have been the most educationally effective approach, needed significant resources to carry out. In fairness, this was always an issue for assessments in large classes pre-COVID-19, but it was brought to the fore during this period.

Workload issues for students became far more visible during COVID-19. Whereas pre-pandemic student workload issues may have been effectively masked by the fact that most courses had a small number of high-stakes assessments in distinct episodes over the term, pandemic teaching brought a shift towards smaller, more frequent assessments. Though these are beneficial to learning, students reported a significant increase in workload. Student stress over assessments does not really scale down with a reduced (or overall small) contribution to final course grade, and the inadvertent outcome for students experiencing these shifts in assessment across a suite of courses they were taking was that an evidence-based assessment strategy inadvertently led to decreased engagement and learning. Course loads, and the workload within individual courses, remains a challenge acting in tension against optimal student learning.

The major challenge for most if not all of these approaches when applied to high-stakes summative assessments was that of academic integrity. COVID-19 essentially de-regulated a highly ordered and controlled invigilation environment for high-stakes summative assessments, and did so effectively overnight. Measures to deal with this change range from an honour-code approach trusting students to self-regulate their academic integrity, all the way to algorithmic remote proctoring applications, which used artificial intelligence algorithms to 'flag' so-called 'events' captured on student webcams during invigilation of the exam (e.g., multiple people in the frame, audio detected etc). These events were then to be reviewed by a course instructor to determine if they merited further action or were innocent interruptions. Such remote proctoring attracted significant attention within and beyond universities. Students found them invasive of privacy and felt they were being surveilled; the efficacy of such applications in preventing integrity violations was questioned (Nigam et al 2021); facial detection algorithms were trained on datasets that were too narrow in terms of ethnicity and gender (Swauger 2020). Yet some academic staff felt these were a necessary deterrent against a rampant growth in cases of academic misconduct, with students facing increased pressure, greater opportunity and easier ability to rationalize academic integrity violations (the elements of the so-called fraud triangle of academic integrity (Choo and Tan 2008)). Several institutions, including ours, passed motions to ban the use of such systems through academic governance channels to their Senates.

The upending of traditional assessment models seemed also to surface anew the well-rehearsed critique of what we are actually doing when we assess students and how the end product—a grade, be it letter, numerical or other classification—drives what and how students learn. A relatively-new adoption into Higher Education (gaining widespread popularity in use and uptake in the 1940s), grades are so deeply ingrained in what we do in our classes we may take it for granted the ways in which they serve useful and multiple purposes to students. A critical examination of the evidence (Schinske and Tanner 2014) suggests they seem to not perform as well as we might think: not appearing to provide effective feedback to inform future effort of students; having the opposite effect on student motivation; creating a competitive classroom environment and often failing to provide reliable information about student learning. Alternative approaches to traditional grading, under the umbrella term of 'ungrading' are borne of a desire to do better by our students, to motivate intrinsic interest in learning, to challenge students to take on difficult learning tasks and to enhance student thinking (Kohn 2006).

On the ground, 'ungrading' approaches look different in different course and disciplinary contexts, but share some common features. There is a focus on instructors providing feedback and students often revising and resubmitting previously-done work, and simple rubrics to gauge levels of attainment with certain outcomes. Practical examples of implementation approaches are becoming more common, to assist instructors in working through the design process of 'how would this work in my course', with details of approaches in, for example, mathematics (Talbert 2022), biology (Leander 2022), and chemistry (Kohn and Blum 2020).

An underlying feature of many of the challenges around assessment during this period was that course design—and assessment design as an integral part of that—has a relatively high inertia, and was challenging to change at the rate demanded by the external circumstances. As we emerge from COVID-19 disruptions, the vast majority of high-stakes assessment is returning to pre-COVID-19 designs, but there remains significant potential for enhancement in this area, even with the practical constraints of class size and resource limitations.

12.7 Conclusion and outlook

Despite the numerous and extended challenges that COVID-19 has brought, the toll it has undoubtedly taken across the teaching and learning community (instructors, support staff, and students), we feel there remains a considerable amount of optimism for the future of large class teaching. Experiences over the last two years have brought hard-earned lessons on how online courses can lead to both innovations and enhancements in some areas of pedagogical practice and enhanced flexibility for learners. Many learners have appreciated the flexibility of 'not everything synchronous and in person' (even if some elements of exactly that were also, simultaneously, keenly missed). From an instructor perspective, we have just undergone the largest scale instructor professional development experiment in higher education, albeit one borne out of necessity. Large class teaching can evolve to incorporate the best of both worlds, balancing the importance of a synchronous, cohort-based learning experience with the added flexibility of a more hybridized approach, with teaching staff that now have first hand experience of navigating this balancing act.

Pre-COVID-19, there was already considerable momentum and progress towards enhanced prioritization of inclusion and well-being in core academic operations. COVID-19 brought obvious challenges to the sense of connection, community, and belonging that students and staff felt during the disruption of regular teaching and learning activities, with consequences for well-being and mental health. These stressors were also distributed unevenly across members of our community, with some of the most vulnerable faced with the most challenging combination of circumstances. At the same time as these obvious challenges, some (learners and staff) felt more productive and connected, with an improved work–life balance. The evolution of approaches to large class course design and delivery need to maintain a focus on inclusive teaching.

Through these disrupted times, we relied on technology more than ever for academic continuity, through wider use of existing applications and adoption of new tools. This brought to the fore considerations of how we support students and staff in this space, as well as concerns around academic integrity and the ethical implications of increased technology use, with a renewed need to go beyond just regulatory compliance around privacy and security when determining if and how to use such tools. Our course design for large classes should (continue to) strive for fluent and appropriate use of technology tools to support specific educational goals.

A period of several cycles of rapid redevelopment and change placed enormous pressures on staff and students operating in process and policy frameworks designed —in most of our institutions—for principally in-person instruction. Staff have had to redesign effectively all courses (often more than once) to accommodate the changing circumstances of the pandemic; support staff have been dealing with both increased volumes and complexity of urgent and novel support requests; students have been the recipients of just-built, untested course designs in challenging contexts. Arrangements for assessments have been brought into particularly sharp focus, along with heightened concerns and increasing volume and complexity of instances of academic misconduct, including the rise of 'contract cheating' third party services. Alternative grading approaches (under the umbrella term of 'ungrading') have gained momentum and may well have planted seeds of ideas for future development.

References

Bates S P 2014 The 21st century educator *Ontario Extend* https://extend.ecampusontario.ca/modules/ (accessed 21 September 2021)

Bruff D 2021 Asynchronous active learning with perusall *Agile Learning* https://derekbruff.org/?p=3717 (accessed 21 November 2021)

Burkholder E W and Wieman C E 2022 Absence of a COVID-induced academic drop in high-school physics learning *Phys. Rev. Phys. Educ. Res.* **18** 023102

Choo F and Tan K 2008 The effect of fraud triangle factors on students' cheating behaviors *Advances in Accounting Education* 9 (Bingley: Emerald Group Publishing Limited) pp 205–20

Freeman S, Eddy S L, McDonough M, Smith M K, Okoroafor N, Jordt H and Wenderoth M P 2014 Active learning increases student performance in science, engineering, and mathematics *Proc. Natl. Acad. Sci.* 111 8410–5 https://www.pnas.org/content/111/23/8410

Hodges C B, Moore S, Lockee B B, Trust T and Bond M A 2020 The difference between emergency remote teaching and online learning *Educause* https://er.educause.edu/articles/2020/3/the-difference-between-emergency-remote-teaching-and-online-learning (accessed 20 August 2020)

Kohn A 2006 The trouble with rubrics *English J* **95** 12–5

Kohn A and Blum S D 2020 *Ungrading: Why Rating Students Undermines Learning (and What to Do Instead)* (Morgantown, WA: West Virginia University Press)

Leander C 2022 Why I ungrade, and a how-to primer *On Teaching and Pie* https://blogs.ubc.ca/celesteleander/tag/ungrading/

Mazur E 2022 How the pandemic changed my teaching: the moral dilemma of going back *MIT Teaching and Learning Lab* https://tll.mit.edu/how-the-pandemic-changed-my-teaching-the-moral-dilemma-of-going-back/ (accessed 1 March 2022

Napierala J, Pilla N, Pichette J and Colyar J 2022 Ontario learning during the COVID-19 pandemic: experiences of Ontario first-year postsecondary students in 2020–21 *Higher Education Quality Council of Ontario* https://heqco.ca/pub/ontario-learning-during-the-covid-19-pandemic-experiences-of-ontario-first-year-postsecondary-students-in-2020-21/ (accessed 15 April 2022)

Nigam A, Pasricha R, Singh T and Churi P 2021 A systematic review on AI-based proctoring systems: past, present and future *Educ. Inform. Technol.* **26** 6421–45

O'Callaghan F V, Neumann D L, Jones L and Creed P A 2017 The use of lecture recordings in higher education: a review of institutional, student, and lecturer issues *Educ. Inform. Technol.* **22** 399–415

Pardo A, Jovanovic J, Dawson S, Gašević D and Mirriahi N 2019 Using learning analytics to scale the provision of personalised feedback *Br. J. Educ. Technol.* **50** 128–38 https://bera-journals.onlinelibrary.wiley.com/doi/abs/10.1111/bjet.12592

Schinske J and Tanner K 2014 Teaching more by grading less (or differently) *CBE—Life Sci. Educ.* **13** 159–66

Swauger S 2020 Our bodies encoded: algorithmic test proctoring in higher education *Hybrid Pedagogy Inc.* https://hybridpedagogy.org/our-bodies-encoded-algorithmic-test-proctoring-in-higher-education/ (accessed 15 March 2021)

Talbert R 2022 Into the ungrading-verse: upgrading to ungrading in an upper-level math course *Grading for Growth* https://gradingforgrowth.com/p/into-the-ungrading-verse

Vigentini L, Kondo E, Samnick K, Liu D, King D and Bridgeman A 2017 Recipes for institutional adoption of a teacher-driven learning analytics tool: case studies from three Australian universities *Proc. ASCILITE2017: 34th Int. Conf. on Innovation, Practice and Research in the Use of Educational Technologies in Tertiary Education* pp 422–32

www.ingramcontent.com/pod-product-compliance
Ingram Content Group UK Ltd.
Pitfield, Milton Keynes, MK11 3LW, UK
UKHW051341160426
5217IPUK00047B/120